Analog Circuit Design for Process
Variation-Resilient Systems-on-a-Chip

Marvin Onabajo · Jose Silva-Martinez

Analog Circuit Design for Process Variation-Resilient Systems-on-a-Chip

 Springer

Marvin Onabajo
Department of Electrical and
   Computer Engineering
409 Dana Research Center
Northeastern University
360 Huntington Avenue
Boston, MA 02115
USA

Jose Silva-Martinez
Department of Electrical and
   Computer Engineering
Texas A&M University
111A Zachry Engineering Building
College Station
TX 77843-3128
USA

ISBN 978-1-4899-9296-3   ISBN 978-1-4614-2296-9 (eBook)
DOI 10.1007/978-1-4614-2296-9
Springer New York Heidelberg Dordrecht London

Printed on acid-free paper

Springer is part of Springer Science+Business Media (www.springer.com)

Preface

Continued improvements of transceiver systems-on-a-chip play a key role in the advancement of mobile telecommunication products as well as wireless systems in medical and remote sensing applications. This book addresses the problems of escalating CMOS process variability and system complexity that diminish the reliability and testability of integrated systems, especially relating to the analog and mixed-signal blocks. Chapter 1 introduces the technical demands and incentives to adopt variation-aware design approaches. The described design techniques and circuit-level attributes are aligned with current built-in testing and self-calibration trends for integrated transceivers, which are explained in Chap. 2. The main attention in this book is on various recent works in which the performances of analog and mixed-signal blocks were enhanced with digitally adjustable elements as well as with automatic analog tuning circuits. To convey the concepts, several case studies are presented that span theoretical aspects and experimental results for variation-aware design approaches related to receiver front-end circuits, baseband filter linearization, and data conversion.

The use of digitally controllable elements to compensate for variations is exemplified with two circuits. First, a distortion cancellation method for operational transconductance amplifiers that enables a third-order intermodulation (IM3) improvement of up to 22 dB is presented in Chap. 3. A transconductance-capacitor lowpass filter with linearized amplifiers is discussed, which was fabricated in a 0.13 μm CMOS process with 1.2 V supply. This filter has a measured IM3 below -70 dB (with 0.2 V peak-to-peak input signal swing) and 54.5 dB dynamic range over its 195 MHz bandwidth. The second example circuit is a 3-bit two-step quantizer with adjustable reference levels, which is the focal point of Chap. 4. This quantizer was designed and fabricated in 0.18 μm CMOS technology as part of a continuous-time $\Sigma\Delta$ analog-to-digital converter system. With 5 mV resolution at a 400 MHz sampling frequency, the quantizer's power dissipation is 24 mW and its die area is 0.4 mm^2.

An alternative to electrical power detectors is explained in Chap. 5 by outlining a strategy for built-in testing of analog circuits with on-chip temperature sensors. Comparisons of an amplifier's measurement results at 1 GHz with the measured

DC voltage output of an on-chip temperature sensor show that the amplifier's power dissipation can be monitored and its 1 dB compression point can be estimated with less than 1 dB error. The sensor has a tunable sensitivity up to 200 mV/mW, a power detection range measured up to 16 mW, and it occupies a die area of 0.012 mm^2 in standard 0.18 μm CMOS technology.

In Chap. 6, an analog calibration technique is discussed to lessen the mismatch between transistors in the differential high-frequency signal path of analog CMOS circuits. The described methodology involves auxiliary transistors that sense the existing mismatch as part of a feedback loop for error minimization. It was assessed by performing statistical Monte Carlo simulations of a differential amplifier and a double-balanced mixer designed in CMOS technologies. Finally, Chap. 7 summarizes the results and conclusions for this calibration case study as well as the other specific design examples in the book.

Marvin Onabajo
Jose Silva-Martinez

Acknowledgments

We would like to extend our sincere appreciation to our colleagues listed below, who made this book possible through collaborations on the described research projects.

Marvin Onabajo, Jose Silva-Martinez

Attenuation-predistortion linearization of operational transconductance amplifiers: Mohamed Mobarak and Edgar Sánchez-Sinencio

Continuous-time lowpass $\Delta\Sigma$ modulator with time-domain quantization and feedback:
Cho-Ying Lu, Venkata Gadde, Yung-Chung Lo, Hsien-Pu Chen, and Vijayaramalingam Periasamy

Observation of RF circuit power and linearity characteristics with on-chip temperature sensors:
Josep Altet, Eduardo Aldrete-Vidrio, Diego Mateo, and Didac Gómez

Contents

Abbreviations

AARCF	Anti-Aliasing Rate Change Filter
AC	Alternating Current
ADC	Analog-to-Digital Converter
ADPLL	All-Digital Phase-Locked Loop
ATE	Automatic Test Equipment
BER	Bit Error Rate
BIT	Built-In Test
BP	Bandpass
BW	Bandwidth
CMFB	Common-Mode Feedback
CMOS	Complementary Metal-Oxide-Semiconductor
CMRR	Common-Mode Rejection Ratio
CT	Continuous-Time
CUT	Circuit Under Test
DAC	Digital-to-Analog Converter
DC	Direct Current
DEM	Dynamic Element Matching
DIBL	Drain-Induced-Barrier-Lowering
DM	Device Mismatch
DNL	Differential Nonlinearity
DR	Dynamic Range
DSB	Double-Sideband
DSP	Digital Signal Processor
DWA	Data Weighted Averaging
ENOB	Effective Number of Bits
ESD	Electrostatic Discharge
EVM	Error Vector Magnitude
FFT	Fast Fourier Transform
FOM	Figures of Merit
HD3	Third-Order Harmonic Distortion
G_m-C	Transconductance-Capacitor

GSM	Global System for Mobile Communications
I/O	Input/Output
I/Q	In-phase/Quadrature-phase
IEEE	Institute of Electrical and Electronics Engineers
IF	Intermediate Frequency
IIP2	Second-Order Intermodulation Intercept Point
IIP3	Third-Order Intermodulation Intercept Point
ILFD	Injection-Locked Frequency Divider
IM2	Second-Order Inter-Modulation
IM3	Third-Order Inter-Modulation
INL	Integral Nonlinearity
IRR	Image Rejection Ratio
IRRX	Image-Reject Receiver
LMS	Least-Mean-Square
LNA	Low-Noise Amplifier
LO	Local Oscillator
LPF	Lowpass Filter
LSB	Least-Significant Bit
MIM	Metal-Insulator-Metal
MOS	Metal-Oxide-Semiconductor
MSB	Most-Significant Bit
NF	Noise Figure
NMOS	N-Channel MOS
NRZ	Non-Return-to-Zero
NSNR	Normalized Signal-to-Noise Ratio
OFDM	Orthogonal Frequency-Division Multiplexing
OSR	Oversampling Ratio
OTA	Operational Transconductance Amplifier
PD	Power Detector
PDA	Personal Digital Assistant
PMOS	P-Channel MOS
PSRR	Power Supply Rejection Ratio
PVT	Process-Voltage-Temperature
PWM	Pulse-Width Modulation
QFN	Quad Flat No-Lead
RC	Resistor-Capacitor
RF	Radio Frequency
RSSI	Received Signal Strength Indicator
RX	Receiver
S/H	Sample-and-Hold
SJNR	Signal-to-Jitter-Noise Ratio
SNDR	Signal-to-Noise-and-Distortion Ratio
SNR	Signal-to-Noise Ratio
SSB	Single-Sideband
TIA	Transimpedance Amplifier

TSMC	Taiwan Semiconductor Manufacturing Company
TX	Transmitter
UMC	United Microelectronics Corporation
UWB	Ultra-Wideband
VCO	Voltage-Controlled Oscillator
WCDMA	Wideband Code Division Multiple Access
Wi-Fi	Wireless Fidelity
WiMAX	Worldwide Interoperability for Microwave Access

About the Authors

Marvin Onabajo is an Assistant Professor in the Electrical and Computer Engineering Department at Northeastern University. He received the B.S. degree (*summa cum laude*) in Electrical Engineering from The University of Texas at Arlington in 2003 as well as the M.S. and Ph.D. degrees in Electrical Engineering from Texas A&M University in 2007 and 2011, respectively.

During his final year at the University of Texas at Arlington, Marvin Onabajo worked in the Analog and Mixed-Signal Integrated Circuit group in affiliation with the National Science Foundation's Research Experiences for Undergraduates program. From 2004 to 2005, he was Electrical Test/Product Engineer at Intel in Hillsboro, Oregon. He was a member of the Analog and Mixed-Signal Center at Texas A&M University from 2005 to 2010, where was engaged in research projects involving analog built-in testing, data converters, and on-chip temperature sensors for thermal monitoring. In the Spring 2011 semester, he worked as a Design Engineering Intern in the Broadband RF/Tuner Development group at Broadcom Corp. in Irvine, California.

Jose Silva-Martinez is currently the Texas Instruments Professor at the Electrical and Computer Engineering Department at Texas A&M University in College Station, Texas. He received the M.Sc. degree from the Instituto Nacional de Astrofísica Optica y Electrónica (INAOE), Puebla, México in 1981, and the Ph.D. degree from the Katholieke Univesiteit Leuven, Leuven, Belgium in 1992.

During his almost 30 years of teaching and research experience at the university level, Jose Silva-Martinez developed several courses in the fields of electronics, circuit analysis, and the design of integrated systems. From 1981 to 1983, he was with the Electrical Engineering Department at INAOE, where he worked on switched-capacitor circuit design. In 1983, he joined the Department of Electrical Engineering, Universidad Autónoma de Puebla, where he pioneered the graduate program in Opto-Electronics in 1992 and remained until 1993. In 1993, he re-joined the Electronics Department at INAOE, serving as the Head of the Electronics Department from May 1995 to December 1998 prior to accepting a faculty position at Texas A&M University in 1999. He has published over 92 journal papers, 150 conference papers, 1 book, and 11 book chapters. His current fields of research are in the design and fabrication of integrated circuits for communication and biomedical applications.

Jose Silva-Martinez has served as Institute of the Electrical and Electronics Engineers' (IEEE) Circuits and Systems Society Vice President in Region 9 (1997–1998), as Associate Editor for the IEEE Transactions on Circuits and Systems (TCAS) Part II (1997–1998, 2002–2003), and as Associate Editor of IEEE TCAS Part I (2004–2005, 2007–2009). He currently serves in the board of editors of six other technical journals. He was the recipient of the 2005 Outstanding Professor Award in the Electrical and Computer Engineering Department at Texas A&M University. He is co-author of the papers that received the 2003 IEEE RF-IC and 2011 IEEE MWCAS Best Student Paper Awards, and co-recipient of the 1990 IEEE ESSCIRC Best Paper Award.

Chapter 1
Introduction

Abstract This chapter introduces the technical challenges associated with increasing variations as a result from scaling down the dimensions of devices fabricated with complementary metal-oxide-semiconductor (CMOS) technology. The incentives for further improvements of variation-aware design and testing techniques are also highlighted. Finally, the chapter contains an overview of the scope and organization of this book.

1.1 Background and Motivation

As rapid progress encompasses the integration of voice, video, and internet connectivity functions into small low-power integrated circuits, portable wireless devices continue to become more prevalent in our lives to the point that many vital situations depend on the reliable operation of the integrated circuits. Consequently, there is an increasing incentive to incorporate self-test and correction features for improved reliability of wireless devices, especially in medical and military applications in which life-saving information is transmitted and received. Even though new technologies allow the design of smaller chips with more functionality, manufacturing process variability and post-production aging effects pose growing challenges for the design, fabrication, and reliability of single-chip mixed-signal systems that are realized with complementary metal-oxide-semiconductor (CMOS) technology in the modern nanometer regime. Consequently, many current research and industry efforts are concentrated on the development of more robust analog and mixed-signal circuits by devising built-in test methodologies that enable digitally-assisted performance tuning.

On the analog circuit level, rising parameter variability is a fundamental contributor to yield and reliability problems. As a result, designing for optimum performance specifications alone is not sufficient anymore. In parallel, it has become critical to

M. Onabajo and J. Silva-Martinez, *Analog Circuit Design for Process Variation-Resilient Systems-on-a-Chip*, DOI: 10.1007/978-1-4614-2296-9_1,
© Springer Science+Business Media New York 2012

improve the on-chip measurement and self-calibration capabilities as well as the testability of single-chip systems during high volume production testing, all in order to increase product yields and to lower the cost of testing. Both yield and cost improvement have been identified as needs in the International Technology Roadmap for Semiconductors [1], giving the incentive for novel built-in test features and alternative test strategies. Additionally, progressive on-chip self-calibration of wireless devices will help to enhance their reliability and allow full utilization of future CMOS technologies with smaller feature sizes despite of increased parameter variations.

Due to high manufacturing volumes, consumer products are a key driving force behind the development of highly integrated chips for wireless communication. For example, the projected global sales of Smartphones are plotted in Fig. 1.1, which is based on the data provided in [2]. The push towards mobile internet and multimedia features has led to ongoing efforts to incorporate additional functionality. At the same time, single-chip transceivers have emerged to perform the analog signal reception/transmission operations and as much digital signal processing on the same chip as possible. This approach has allowed to reduce product dimensions and production cost. Nowadays, cell phones have fewer chips on the printed circuit board (Fig. 1.2), but the complexity of those chips causes significant design and testing complications. In the case of integrated transceivers, the demand to support multiple communication standards has created design issues related to more stringent linearity requirements for the broadband radio frequency (RF) front-end circuits, reconfigurability of many blocks along the transmit/receive chains, interference avoidance among circuits, minimization of total power consumption, and other aspects. Within the scope this book is that RF system performance monitoring is becoming significantly more important and difficult with the trend towards increasing integration and power densities in single-chip systems fabricated with modern CMOS technologies. On-chip electrical power detectors are commonly used to monitor and optimize the dynamic range of RF systems through measurements and controlled amplifications in RF front-ends. However, the adverse effects from parasitic input capacitances of electrical detectors become more detrimental at higher frequencies. Non-invasive temperature sensors for RF power detection offer an attractive alternative to conventional power detectors, as shown by the investigations presented in this book.

1.2 Book Scope and Organization

Contemporary CMOS technologies have offered progress with respect to circuit properties such as smaller device dimensions, better high-frequency operation, and power efficiency. But, analog designers in particular face various technology-related drawbacks associated with newer technologies, such as signal swing limitations due to decreased supply voltage and gain reduction due to lower transistor output resistance. Other major disadvantages, which are elaborated in

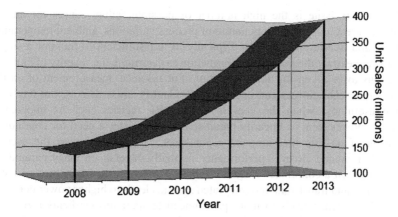

Fig. 1.1 Smartphone market trend

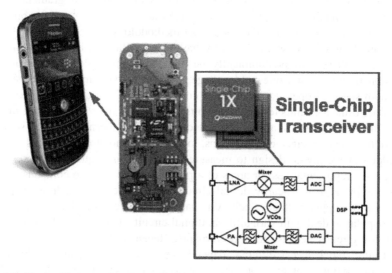

Fig. 1.2 Single-chip transceiver in a cell phone

Chap. 2, are worsening process variations and intra-die device mismatches. These have a strong impact on the product yield and reliability, translating into manufacturing cost as well as critical risk factors in medical or military applications. Variations and circuit sensitivity to environmental conditions such as temperature changes and interference from other nearby circuits are becoming more problematic as the complexity of integrated systems increases. In this book, special attention is given to augmentations of analog and mixed-signal circuits in response to the emerging variability problems and system-level calibration approaches concerning current and future CMOS technologies.

An intricate issue is the high number of possible failure causes for analog circuits as a result of the random nature of process variations, ambient temperature changes, and interference signals. Typically, it is insufficient to monitor a single quantity and extract the necessary information to determine the severity of faults or the actions to be taken for their correction. For instance, measurement of an RF circuit's quiescent current can be helpful to identify gross defects, but has very limited usefulness when the goal is to tune RF metrics such as impedance matching, power gain, or linearity parameters. This creates a need for continuous expansion of on-chip measurement capabilities, especially because the acceptability of an analog circuit's performance normally depends on many parameters that can take on a continuous range of values. Moreover, the integration of more functionality and transistors into integrated systems leads to higher power densities on the chips, which leads to more pronounced temperature gradients and interference between circuits due to thermal coupling. A temperature sensing strategy is introduced in Chap. 5 to provide alternative means for on-chip measurements of RF characteristics and to increase the observability of temperature gradients. The chapter also contains descriptions of a temperature sensor topology for built-in testing of analog circuits and the simulation methodology for its design.

A digital circuit whose functionality has been verified during the characterization test phase will predominantly be affected by process variation of the transition frequency and threshold voltage, which will have main effects on the maximum speed of operation and power consumption. This eases the determination of performance limits for digital circuits by verifying their logic outputs or the output of test structures at the mandated speed. As alternative for test cost reductions or performance optimizations, local process monitoring devices can be embedded in the layout design to measure the transition frequency or threshold voltages (as representatives for areas of a partitioned die), and to compensate for variations by adjusting nearby digital circuits through features such as adaptive body bias or supply voltage. Such systematic approaches have become increasingly popular to deal with variability in digital circuits, but they are less effective for analog circuits because their performance depends on more parameters and each analog block has a different dependence on a given parameter. For that reason, the design strategies for robust analog circuits tend to be tailored to the circuit type or even its specific topology.

The approach taken in this book is to present examples of circuits and their features that alleviate the effects of process variations. With adaptations, the presented methodologies can be extended to similar analog circuits. In particular, the use of digitally programmable circuit elements or bias conditions will be emphasized and related to the compatibility of individual blocks with emerging system-level self-calibration strategies. The first example to be discussed in Chap. 3 is the linearization of transconductance amplifiers in broadband filter applications. Chapter 4 describes another case study, which is a 3-bit quantizer that was designed for continuous-time $\Sigma\Delta$ analog-to-digital converters. Chapter 5 introduces a strategy to utilize differential temperature sensors as on-chip RF power detectors for built-in testing. Next, a general technique to reduce the mismatch between transistors is described in Chap. 6, in which it is applied to differential pair transistors of a wide bandwidth amplifier and the switching transistors of a double-balanced mixer. Chapter 7 provides summaries and conclusions for the deliberated topics. The following subsections give a more detailed overview of the focal points in this book.

1.2.1 Linearization Scheme for Transconductance Amplifiers

Operational transconductance amplifiers (OTAs) are elements of transconductance-capacitor (G_m-C) filters found in many wireless receivers and continuous-time $\Sigma\Delta$ analog-to-digital converters. Thus, OTA performance and dependability improvements manifest themselves in system-level enhancements of communication circuits and sensor signal conditioning circuits. The push towards wider bandwidths in these applications mandates OTA designs with progressively better linearity at higher frequencies. Towards this end, an architectural solution is described in Chap. 3 that can be applied to diverse circuit-level OTA configurations. Effective linearization over a wide frequency range demands a mechanism to correct for high-frequency effects and process variations. Accordingly, digital programmability was realized to ensure high linearity and compatibility with modern CMOS technologies.

The linearization technique utilizes two matched OTAs to cancel output harmonic distortion components, creating a robust architecture. Compensation for process variations and frequency-dependent distortion based on Volterra series analysis is achieved by employing a delay equalization scheme with on-chip programmable resistors. An OTA design with this broadband linearization method has third-order inter-modulation (IM3) distortion better than -74 dB up to 350 MHz with $0.2V_{p-p}$ input, 70 dB signal-to-noise ratio (SNR) in 1 MHz bandwidth, and 5.2 mW power consumption. The distortion-cancellation technique enables an IM3 improvement of up to 22 dB compared to a commensurate OTA without linearization. A proof-of-concept lowpass filter with the linearized OTAs has a measured IM3 below -70 dB and 54.5 dB dynamic range over its 195 MHz bandwidth. The standalone OTAs and the filter were fabricated on a 0.13 μm CMOS test chip with 1.2 V supply.

1.2.2 Process Variation-Aware Quantization

Future wireless devices will require extensive connectivity to accommodate several services, which means that the receivers must cover broader frequency bands. Therefore, on-chip analog-to-digital converters (ADCs) in multi-standard receivers not only demand increased signal-to-quantization-noise-ratio, but also more bandwidth for the conversion of the analog signal into the digital domain. Chapter 4 briefly introduces a lowpass continuous-time $\Sigma\Delta$ ADC architecture that was developed for next generation broadband receiver applications. Rather than using multiple signal levels, a multi-bit digital-to-analog converter (DAC) realization based on a feedback signal with time-varying pulse duration was employed. This approach alleviates nonlinearity problems associated with typical multi-bit DACs. Chapter 4 of this book describes the corresponding 3-bit quantizer architecture with multi-phase clocking. The reference levels for the quantizer are adjustable to compensate for process variations after fabrication if the application necessitates fine resolution. Designed with 5 mV resolution at a 400 MHz sampling frequency, the quantizer power dissipation is 24 mW and its die area with auxiliary logic circuitry and routing is 0.4 mm^2. With embedded quantizer, the 5th-order $\Sigma\Delta$ ADC achieves a measured peak signal-to-noise-and-distortion ratio (SNDR) of 67.7 dB in 25 MHz bandwidth. Fabricated in a standard 0.18 μm CMOS technology, the ADC consumes a total of 48 mW with a 1.8 V supply, and occupies 2.6 mm^2 die area.

1.2.3 Non-Invasive On-Chip Measurement of Thermal Gradients and RF Power

One aspect of designing robust analog and mixed-signal circuits in wireless products is the inclusion of on-chip monitors that can determine whether device performance parameters are within an acceptable range or whether a detrimental shift has occurred due to effects from aging, temperature variations, interfering signals, or other conditions. This information can then be incorporated into self-calibration schemes that tune circuit blocks to restore satisfactory functionality. A part of this book is directed towards the conception of a practical monitoring strategy with differential temperature sensors having high sensitivity and accuracy for the measurement of on-chip temperature gradients. Due to thermal coupling, the temperature in the vicinity of a device depends on its power dissipation, and this relation can be exploited for testing purposes.

In Chap. 5, a design methodology is presented which aims at the extraction of RF circuit performance characteristics from the DC output of an on-chip temperature sensor. Any RF input signal can be applied to excite the circuit under examination because only dissipated power levels are measured, which makes this approach attractive for online thermal monitoring and built-in test scenarios. A fully-differential sensor topology is introduced that has been specifically designed for this method by constructing it with a wide dynamic range, programmable sensitivity to DC and RF power dissipation, as well as compatibility with CMOS technology. Furthermore, a procedure is outlined to model the local electro-thermal coupling between heat sources and the sensor, which is used to define the temperature sensor's specifications as well as to predict the thermal signature of the circuit under test.

Chapter 5 also summarizes the experimental results obtained from a prototype chip with an RF amplifier and temperature sensor fabricated in a conventional 0.18 μm CMOS technology. The temperature-sensing concepts were validated by correlating RF measurements at 1 GHz with the measured DC voltage output of the on-chip sensor as well as with simulation results, demonstrating that the RF power dissipation can be monitored and the 1 dB compression point can be estimated with less than 1 dB error. The sensor circuitry occupies a die area of 0.012 mm^2, which can be shared when several on-chip locations are observed by placement of multiple temperature-sensing parasitic bipolar devices.

1.2.4 Analog Calibration for Transistor Mismatch Reduction

An analog calibration technique is described in Chap. 6, which aims at lessening the mismatch between transistors in the differential high-frequency signal path of analog CMOS circuits. It can be applied for offset reduction in high-speed amplifiers and comparators in which short-channel devices are utilized to minimize bandwidth reduction from parasitic capacitances. In general, it is suitable for RF applications in which direct matching of the transistors is undesired because sophisticated layout practices would increase the coupling between the high-frequency paths. The methodology involves auxiliary devices that sense the existing mismatch as part of a feedback loop for error minimization. This technique is demonstrated in Chap. 6 with a differential amplifier having a loaded gain and −3 dB frequency of 13 dB and 2.14 GHz. This amplifier was designed in 90 nm CMOS technology with a 1.2 V supply. Monte Carlo simulations indicate that the 4.17 mV standard deviation of the amplifier's anticipated input-referred offset voltage improves to 0.76–1.29 mV with the mismatch reduction loop, which is contingent on the layout configuration of the mismatch-sensing transistors.

Chapter 6 also provides a second application example for the analog mismatch reduction loop, which is to enhance the matching between the switching transistors in a double-balanced CMOS mixer. Simulation results show that this scheme improves the mixer's IIP2 by 5 dB while having negligible impact on other performance parameters with the exception of 30% higher power due to the dissipation in the calibration circuitry. The calibration method helps to compensate for the large process variations of the mixer transistors that are biased with small currents in the subthreshold region. As a result, the power consumption of the presented mixer is still more than six times lower than that of conventional downconversion mixers using saturation region bias, whereas its specifications are similar to the state of the art.

References

1. *International Technology Roadmap for Semiconductors, Test and Test Equipment*, 2009 edn. Available: http://public.itrs.net/reports.html
2. S. Menon, C.L. Horney, Smartphone & Chip Market Opportunities, Market research report no. 9010, Forward Concepts Co., Available: http://fwdconcepts.com/Smartphones. 5 Feb 2009

Chapter 2
Process Variation Challenges and Solutions Approaches

Abstract The technical and economic impacts of worsening process variations and intra-die device mismatches are elaborated in this chapter, especially with regards to product yield, reliability, and manufacturing cost. This introduction is followed by a survey of diverse built-in testing and calibration approaches aimed at enhancing performance, yield, and reliability in the presence of variations.

2.1 Current Trends

2.1.1 The Impact of Rising Process Variations

Most semiconductor product improvements over the past decades are direct or indirect consequences of the perpetual shrinking of devices and circuits, allowing performance enhancements at lower fabrication cost. A paralleling trend is that process variations and intra-die variability increase with each technology node. Since most high-performance analog circuits depend on matched devices and differential signal paths, this trend has begun to diminish yields and reliabilities of chip designs. Fundamentally, the problem is that parameters of devices on the same die show increasing intra-die variations, thereby exhibiting different characteristics. For example, Table 2.1 displays the evolution of the typical transistor threshold voltage standard deviation $\sigma\{V_{Th}\}$ normalized by the threshold voltage (V_{Th}) for several technologies, as reported in [1]. Also notice that V_{Th} exhibits further dependence on gate length variations through the drain-induced-barrier-lowering (DIBL) effect under large drain-source voltage bias conditions, as demonstrated by the characterization in [2] using 65 nm technology. Since DIBL worsens as the channel is scaled down, this additional impact on threshold voltage variations can be assumed to be even stronger beyond the 65 nm technology node.

M. Onabajo and J. Silva-Martinez, *Analog Circuit Design for Process Variation-Resilient Systems-on-a-Chip*, DOI: 10.1007/978-1-4614-2296-9_2,
© Springer Science+Business Media New York 2012

Table 2.1 Intra-die variability vs. CMOS technology node

Technology node	250 nm (%)	180 nm (%)	130 nm (%)	90 nm (%)	65 nm (%)	45 nm (%)
$\sigma\{V_{Th}\}/V_{Th}$	4.7	5.8	8.2	9.3	10.7	16

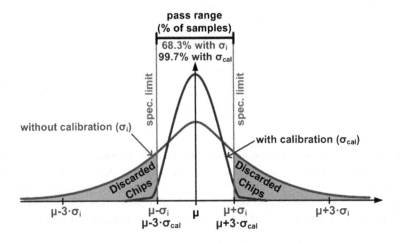

Fig. 2.1 Specification variation impact on the fraction of discarded chips

A direct consequence of device parameter variations is a decrease in production yields because block-level and system-level parameters will show a corresponding increase in variations. This relationship between variations and yield can be inferred from the visualization in Fig. 2.1, where the Gaussian distribution of a specification with a standard deviation σ around the mean value μ is shown together with the specification limits ($\pm 3\sigma$ in this example). For standalone analog circuits, parameters such as gain may have an upper and/or lower specification limit, and the samples that exceed the limit(s) during production testing must be discarded. Guardbands are often defined to account for measurement uncertainties by following procedures such as repeating the same test or performing other more comprehensive tests to determine whether the part can be sold to customers, which incurs additional test cost in a manufacturing environment.

An important observation from Fig. 2.1 is that an increase of variation (σ) widens the Gaussian distribution, which leads to a higher percentage of parts that fall within the highlighted ranges that require them to be scrapped or retested. Clearly, there is a direct relationship between the amount of process variations and production cost due to low yields. In the case of wireless mixed-signal integrated systems, the trend towards increasing integration and complexity has also been paralleled by technical challenges and rising cost of testing, which can amount up to 40–50% of the total manufacturing cost [3, 4]. As a consequence, built-in self-test, design-for-test, and design-for-manufacturability methods for analog and mixed-signal circuits have received growing attention over the past years.

Fig. 2.2 Process corner-based vs. 3σ design approaches

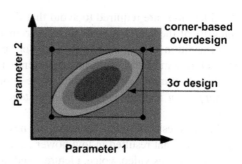

2.1.2 Circuit and System Design Tendencies

System complexities and process variations raise the importance of considering testability early in the design phase to avoid technical complications and time-to-market delays in the pre-production phase as well as test cost reduction during the production phase. Worst-case process corner models have been used extensively to account for variations during the design of analog circuits. But more recently, a paradigm shift towards the use of statistical models and Monte Carlo simulations has occurred. One of the main reasons for this development is that corner-based design easily results in too pessimistic designs [5], which is evident in Fig. 2.2. In this figure, the x-axis and y-axis represent the ranges over which two parameters can vary, and the area inside the ellipse indicates the combined range in which the 3σ limits are met. This region can be predicted with statistical Monte Carlo simulations for yield estimation. On the other hand, the area outside of the elliptical design space corresponds to design implementations that meet the specifications, but are over designed. This means that "investments" of area, power, or trade-offs with other parameters are made in order to allow acceptable performance despite of increased deviations of the two parameters from their nominal values. The rectangular region between the combination of the four worst corner cases of the two parameters includes over design space, implying that it involves costly performance or parameter trade-offs. This economic reason and the availability of more efficient computational tools have created a trend towards statistical yield optimizations rather than corner-based design [5].

Defect densities on wafers become worse in newer technologies and production yields decrease with increased chip size [6]. Self-test and self-repair schemes for digital circuits have been routinely incorporated into products for a long time, especially since on-chip verification of logic blocks and repair with redundant circuitry do not require analog instrumentation resources. The inclusion of scan chains gives easy access to internal digital circuitry through a minimal number of pins during production testing. Similarly, the standardized mixed-signal test bus (IEEE Std. 1149.4) has been developed to improve the testability of analog blocks by allowing better observation of internal nodes. Nowadays, the use of analog test buses within single-chip systems is feasible in the industry, but significant design considerations are required to avoid that the interface circuitry does not affect the integrity of the analog signals or measurements [7].

In addition to the underlying variation and defect issues on the device level, several system-level and technology trends impair the testability and manufacturability of integrated circuits for mobile applications:

Support of multiple communication standards and more features on low-power chips

The wireless communication industry has experienced phenomenal growth in the past decade that resulted in low-power handheld devices with multi-purpose functionality such as video, voice, pictures, and internet access. The wireless local-area networks for laptops, desktops, and personal digital assistants (PDAs) include standards like Bluetooth, WiFi, IEEE 802.16, WiMAX, Ultra-Wideband (UWB), and GPS. Most relevant services for handheld devices range from 470 MHz to almost 11 GHz. The main technical challenge is the co-existence of wireless devices, which results in signal interference. This can be solved if more linear high-performance analog receiver front-ends are available to tolerate and filter out high-power interfering signals without saturation of the analog blocks due to high signal power levels. Further filtering and channel selection can be performed in the digital domain when the signal integrity is maintained by the processing through unsaturated highly-linear analog blocks. Support of multiple communication standards requires chips with more circuitry and complexity, which makes them less testable in the production stage because of limited access to internal nodes, interactions between blocks, and a higher number of test cases to verify functionality. Systems with more subcomponents are more likely to fail, which is another reason why yields of integrated receivers, transmitters, and transceivers are on the decline. Simultaneously, the processing of broadband signals in their front-ends mandates high-performance analog circuits, which in many cases requires continued circuit-level innovations for on-chip self-calibration to tune for optimum performance.

Process technology optimizations for digital circuits create analog design challenges

The main advantages of device scaling with CMOS technology are improved performance at higher frequencies, reduced power consumption, and increased levels of integration. Those benefits are particularly aiding the development of digital circuits and systems. With regards to analog circuits, deep-submicron technology scaling progress comes together with adverse effects such as reduced gains from lower transistor output impedances, design with limited voltage headroom, higher flicker noise levels, and reduced transistor linearity. Larger variability of parameters is caused by physical and fabrication limitations such as under-etching uncertainties, variations of effective transistor dimensions, severe channel length modulation due to higher electric fields, and channel dopant fluctuations. Interestingly, the random dopant fluctuations have reached a severity that can lead to significant threshold voltage mismatch in neighboring devices at the 65 nm node [8]. Additional reliability concerns arise from the restricted power that transistors can supply to the load without exceeding the low breakdown voltage of the deep submicron devices. Furthermore, digital CMOS processes often do not provide high-quality passive devices required for conventional high-performance analog designs. For example, metal-insulator-metal (MIM) capacitors, high-resistivity polysilicon resistors, or well-characterized inductor models might not be available in a digital process, forcing analog designers to get by with metal-oxide-semiconductor (MOS) capacitors and standard polysilicon resistors. Both of these have higher parasitic capacitances to the substrate than the equal-valued MIM capacitors or high-resistivity polysilicon resistors. Scaling down transistors permits more digital functionality and memory on a single chip, but with less reliability especially for analog signal processing.

2.2 System Perspective on Transceiver Built-In Testing and Self-Calibration

The concepts and examples presented in this book are all involving circuit blocks which are found in conventional transceivers within mobile wireless devices. While equipping the circuit blocks with built-in test (BIT) and self-calibration features to compensate for variations, it is important to keep their role as part of the system in mind because of the interaction between blocks and the overall goal to optimize system-level performance specifications such as bit error rate (BER) or error vector magnitude (EVM). In general, the self-calibration challenge can be divided into two parts: one is to add tunability and controllability capabilities in the individual blocks, and the other one is to devise comprehensive system-level calibration algorithms in a digital signal processing unit. The former task is the focus of this book, but the existing approaches for the latter task will be briefly discussed next and when applicable throughout the book.

BIT strategies for transceivers vary tremendously depending on the transceiver architecture, communication standard, available on-chip measurement and computation resources, the production volume, and whether the BIT is designed for production testing (quality control) or on-line self-calibration (reliability) during the life time of the chip. Consequently, most BITs involve a mix of analog and digital blocks, on-chip and off-chip measurement devices, long calibration routines at start-up, and shorter periodic or on-line calibration. Generally, a trend has emerged to combine techniques for verification of complex mixed-signal transceivers implemented as single chips. Nevertheless, the BIT approaches can be grouped into a few rough high-level categories that represent the different design philosophies in academia and the industry. In the following overview, a few example cases will be discussed to highlight the distinctive characteristics of methods that can be broadly classified into the categories below.

- Digital correction and calibration (digitally assisted)
- Analog measurements and tuning
- Loopback testing
- Combined digital performance monitoring and analog compensation
- Combined digital monitoring, analog measurements, and analog compensation

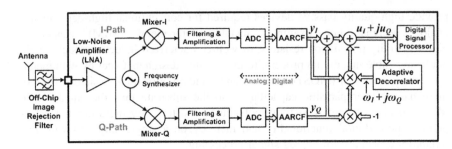

Fig. 2.3 Receiver with digital I/Q mismatch compensation

2.2.1 Digital Correction and Calibration

Digital BIT approaches involve measurements and compensation techniques that are realized in the digital baseband processor of the transceiver. They are suitable for parameters that are observable and traceable in the digital domain, such as slowly drifting DC offsets or mismatch between the in-phase (I) and quadrature-phase (Q) paths in the front-end. Generally, digital methods have the advantage of high precision when sufficient computational resources are available. They are also very attractive for on-line calibration schemes that run in the background.

Digital I/Q mismatch compensation is a widely used method that involves digital measurement and compensation of the I/Q gain and phase mismatches in the analog front-end circuitry. For example, the work in [9] presents a scheme that runs during start-up or in a dedicated calibration mode to ensure acceptable performance of a low-IF receiver even with up to 10% gain and 10° phase imbalance in the analog front-end. On-line digital I/Q compensation techniques have also been reported, such as [10], in which the training symbols that are standard in orthogonal frequency-division multiplexing (OFDM) transmissions are exploited for background I/Q calibration. It was also demonstrated in [10] how digital I/Q compensation relaxes the overall signal-to-noise ratio (SNR) requirements in the receiver chain because I/Q imbalance directly affects the SNR and thereby degrades the bit error rate (BER). In the OFDM receiver example presented in [10], the digital calibration allowed to improve the tolerance to I/Q imbalances from 1%-gain/1°-phase to 10%-gain/10°-phase.

Digital I/Q calibration is widely used in the industry. An example is the work from Texas Instruments describing a low-IF GSM receiver in 90 nm CMOS technology [11]. This receiver utilizes an adaptive filter that obtains the mismatch information from on-line I/Q correlations, for which the modified block diagram from [11] is displayed in Fig. 2.3. The interesting part of the block diagram is the adaptive decorrelator after the analog-to-digital converter (ADC) and anti-aliasing rate change filter (AARCF). In the digital domain, gain mismatch appears as difference in the auto-correlation between I and Q paths, while phase mismatch appears as nonzero cross-correlation between I and Q. The authors use an algorithm that takes advantage of the aforementioned relationships by implementing an adaptive decorrelator which attempts to minimize the auto-correlation and the cross-correlation between I and Q outputs (y_I, y_Q). This is done by adjusting the correction coefficients:

$$\omega_{I(n+1)} = \omega_{I(n)} + \mu \cdot \left[u_{I(n)} \cdot u_{I(n)} - u_{Q(n)} \cdot u_{Q(n)} \right], \omega_{Q(n+1)}$$
$$= \omega_{Q(n)} + 2\mu \cdot u_{I(n)} \cdot u_{Q(n)} \qquad (2.1)$$

where μ is the adaptation step size which is inversely proportional to the signal energy. Thus, periodic training sequences are required with this scheme. Depending on process-voltage-temperature (PVT) variations, 15–30 dB image rejection ratio (IRR) improvement has been demonstrated in practice with phase mismatch <1° and amplitude mismatch <10% in [11] with a settling time in the range of 3–4 ms. This settling time is lengthy compared to analog tuning approaches that can be as short as a few microseconds [12], which becomes important in production testing situations because any adjustments for different test conditions in the front-end (different gain settings, channel, etc.) would require 3–4 ms idle time for digital I/Q calibration before the BER test can begin. On the other hand, settling times of analog tuning schemes depend on the loop bandwidth, which can be designed in the megahertz range to achieve settling times in the microseconds regime. Hence, analog I/Q tuning approaches would fill the niche of situations that require fast convergence.

The incentive for using a digital BIT technique is high when the circuit under test itself has digital features. An example is the BIT of a transmitter in [13] that includes an all-digital phase-locked loop (ADPLL). In that case, the error signal of the ADPLL is already in the digital domain, allowing to monitor failures and the center frequency drift of the digitally controlled oscillator. Furthermore, the authors of [13] state that digital filtering and spectral estimation can be used to monitor and adjust the phase noise transfer function.

2.2.2 Analog Measurements and Tuning

The analog equivalent to the digital I/Q imbalance calibration scheme has been proposed and demonstrated for image-reject receiver (IRRX) architectures. A simplified block diagram of such a BIT is displayed in Fig. 2.4, which is representing the work from [14]. In an IRRX, the down-conversion scheme with two mixing stages suppresses the image signal at the second intermediate frequency output Out(f_{IF2}), which avoids the need for an external image-rejection filter. The quality of the image-rejection is typically expressed with the image-rejection ratio (IRR) that depends on the I/Q amplitude mismatch (ΔA) and phase mismatch ($\Delta \theta$):

$$IRR_{(dB)} \approx 10 \cdot \log \left((1/4) \cdot [(\Delta \theta)^2 + (\Delta A/A)^2] \right) \qquad (2.2)$$

In practice, the IRR is normally limited to 25–40 dB due to mismatches, even though almost 60 dB are required for acceptable BER performance. In [14], a

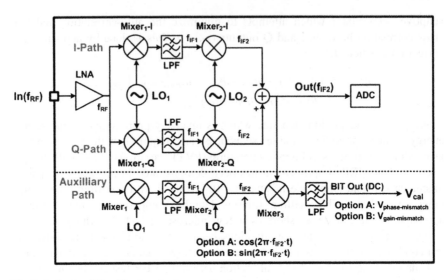

Fig. 2.4 Analog I/Q calibration for image-rejection receivers

purely analog calibration scheme was implemented with the auxiliary path shown in Fig. 2.4. This path contains the duplicate mixing operations as in the main path with the exception that the output signal at the second intermediate frequency (f_{IF2}) can be of the form $\cos(2\pi \cdot f_{IF2} \cdot t)$ or $\sin(2\pi \cdot f_{IF2} \cdot t)$, depending on which phases of the two local oscillators (LO_1, LO_2) are routed to the auxiliary mixers. Finally, $mixer_3$ correlates the signals from the two paths to extract the I/Q mismatch information contained in the DC component after the lowpass filter (LPF). This analog DC voltage (V_{cal}) can be directly used to tune the bias voltages of analog circuits for mismatch compensation, resulting in high IRR (e.g. 57 dB in [14]). A similar automatic IRR calibration with analog mixers, variable phase shifter, and gain tuning has been realized in [15] with an IRR of 59 dB.

A benefit with analog tuning is that the bias conditions of the analog blocks under calibration are controlled and less affected by PVT variations due to the correcting action of the loops, thereby allowing higher yields as a result of automatic correction in the analog front-end. However, the power and area consumption of the BIT circuitry is the main trade-off. In addition, the BIT circuits themselves have to be designed robustly to avoid failures, making the implementation more challenging and invasive than digital calibration schemes. Efforts for the analog approach are generally more justified in transceivers that have few on-chip digital resources and in scenarios that require fast automatic correction. For example, the IRR calibration in [15] can be used on-line with a settling time that depends on the bandwidth of the analog control loops rather than convergence of digital algorithms which take several milliseconds as in [11]. Another fast analog calibration method with a convergence time in the microseconds regime is described in [12].

Fig. 2.5 BIT with analog
instrumentation along the
signal path

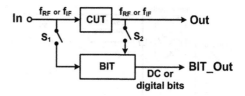

Instead of using a system-level test strategy, it has been more popular to extract information from each block in the analog front-end for characterization or tuning of the individual block, which is visualized in Fig. 2.5. The circuit under test (CUT) represents a block in the RF front-end or analog baseband that can be connected to a BIT circuit in test mode by closing the two switches S_1 and S_2. In [16] for instance, a low-noise amplifier (LNA) was tested with a BIT block containing a test amplifier and two power detectors to measure input impedance, gain, noise figure, input return loss, and output SNR of the LNA. This approach has the advantage that the fault location/cause can be identified clearly and that the DC or digital outputs of the BIT circuits can be used to recover from certain failure modes. High-frequency RF front-ends have been targeted in particular with dedicated design of BIT circuits because gain, impedance matching, and linearity performances are very sensitive to variations. Furthermore, direct signal digitization is not feasible at high frequencies, eliminating many digital compensation schemes. Hence, several RF block-level measurement approaches involve power or amplitude detectors along the signal path [17–20].

Self-calibration of impedance matching for an LNA at the input of the receiver chain as done in [21] also requires on-chip analog sensing circuitry, especially to achieve a short calibration time such as the 30 µs reported in [21]. An alternative proposition to monitor individual blocks in the signal path was made in [22], in which the transient supply currents of the CUTs are monitored with the BIT circuitry by placing small series resistors in the power supply lines. However, a clear disadvantage with any block-level measurement is that the BIT circuitry is connected to the CUT and therefore must be designed carefully to avoid impact on performance. But, some degradation due to loading effects from BIT circuitry must usually be tolerated. Furthermore, switches in or along the signal path are undesired due to their added noise, power losses and signal feed through from finite isolation, particularly at RF frequencies.

Though with less accuracy than off-chip measurement equipment, efforts have also been made to mimic conventional instrumentation such as spectrum analyzers [23, 24] on the chip with sufficient accuracy for BIT applications. In [23] for example, the analyzer with a frequency range of 33 MHz–3 GHz could cover the entire signal paths of many wireless transceivers in handheld consumer products. A multiplexer could be used to selectively route a test input at a time to one spectrum analyzer, but the on-chip measurement circuitry still takes up large area and significant power that might not be permissible in certain applications. For example the analyzer in [23] consumes 0.384 mm^2 and more than 20 mW.

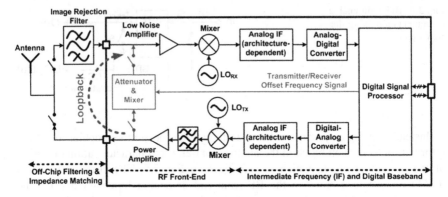

Fig. 2.6 Generalized transceiver block diagram with loopback

2.2.3 Loopback Testing

Loopback testing is a system-level BIT technique in which the BER is monitored in the digital baseband [25]. It allows simultaneous verification of the analog and digital transceiver blocks (Fig. 2.6) with a low-frequency digital input signal applied to the baseband subsection of the transmitter. This up-converted signal is routed from the transmitter (TX) output to the receiver (RX) input via a loopback connection [26]. After down-conversion and digitization in the RX, the received bitstream is analyzed in the digital baseband processor to determine the BER. Attenuation and frequency translation with a mixer are required in the loopback block to maintain signal integrity and to ensure that the power levels during testing are comparable to normal operation. If the communication standard does not require frequency translation between TX and RX, then only the attenuator is required. In any case, the overhead of the BIT circuitry is below 10% of the complete transceiver, which is efficient. However, the loopback BIT cannot be executed on-line; it requires a dedicated test mode during production testing or self-checks during times when the transceiver is idle.

The main benefit of the loopback technique is that a BER test is the most important metric, which is only low when all components function properly. This property makes loopback very attractive for fast pass/fail production testing and quick self-checks during in-field use, especially when few or no off-chip test resources are available. For example, a loopback test for the on-wafer production test stage was presented in [27].

A drawback of early loopback implementations is the lack of information regarding failure causes and fault locations. In response, one proposed variant [28] involves more computations in the digital baseband processor to determine the spectral content of the received bits and to use the data for estimation of receiver/ transmitter nonlinearity specifications. Alternatively, power detectors could be placed at critical nodes to extract block-level gain and 1 dB compression point measurements. Or, similarly, statistical sampling blocks were placed along the

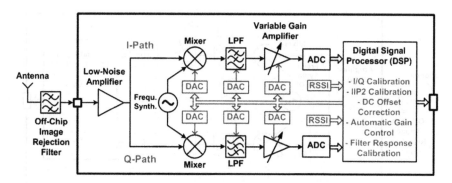

Fig. 2.7 Transceiver with digital monitoring and tuning of analog blocks

signal path in [29]. These blocks produce digital bitstreams for analysis of fault locations. In general, inclusion of auxiliary circuitry during a loopback test increases the observability of faults, but with the associated trade-offs that have been discussed for on-chip measurement circuitry in Sect. 2.2.2.

2.2.4 Digital Performance Monitoring with Analog Compensation

A BIT approach for complex transceiver chips that has become increasingly popular in recent years is depicted in Fig. 2.7. It incorporates accurate digital monitoring and I/Q mismatch correction in the baseband processors as well as a few analog observables such as outputs from received signal strength indicators (RSSIs) or DC control voltages of blocks that give some insights into their operating conditions. A significant aspect is that many analog bias voltages for RF front-end and baseband circuits are generated with digital-to-analog converters (DACs). These DACs are utilized for coarse adjustments at start-up in order to compensate for PVT variations. They also reduce DC offsets in the analog circuits to prevent saturation of internal nodes due to large gains in the receiver. Thus, more mismatches can be tolerated because of the capability to counteract them.

Combined digital monitoring/calibration with analog compensation DACs has been reported in publications describing industrial transceivers. Some examples are:

- Single-chip GSM/WCDMA transceiver in 90 nm CMOS [30], Freescale, 2009

  – DC offset, I/Q gain & phase, IIP2 calibration in the digital signal processor
  – 6-bit DACs for analog compensation

- 2.4 GHz Bluetooth Radio in 0.35 μm CMOS [31], Broadcom, 2005

  – Bias networks with digital settings for LNA, mixer, filter

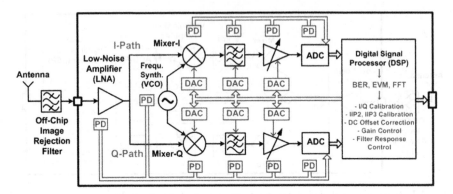

Fig. 2.8 Transceiver with digital monitoring, analog measurements, and tuning

- Tuning patent (US 7,149,488 B2); RSSIs & digital block-level bias trimming

- 2.4 GHz WLAN transceiver in 0.25 μm CMOS [32], MuChip, 2005

  - Baseband I/Q gain and phase calibration
  - Extra analog mixer & peak detector

- 5.15–5.825 GHz WLAN transceiver in 0.18 μm CMOS [33], Athena, 2003

  - Digital I/Q mismatch correction
  - Multiple internal loopback switches for self-calibration in test mode
  - 8-bit DACs for DC offset minimization after mixers and filters

2.2.5 Combined Digital Monitoring, Analog Measurements, and Tuning

The circuit-level research projects discussed in the following sections are based on the hybrid analog/digital approach in the previous subsection. One goal is to improve fault observability and calibration effectiveness by adding more measurement circuitry in the analog segments to provide data that can become part of the system-level calibration routine. Information from measurements can be used for block-level tuning prioritizations and optimizations, leading to shorter start-up routines and convergence times of algorithms. Figure 2.8 portrays the envisioned transceiver with enhanced analog measurements, where power detectors (PD) measure gains along the analog chain [17–20]. Power gain and linearity measurements through temperature sensing are explored in Chap. 5. In contrast to conventional power detectors, temperature sensors do not physically come in contact with the CUT and thus avoid loading effects.

Another aspect of comprehensive system-level self-calibration is that the analog circuits must have tunable or programmable elements, meaning that "knobs" to adjust performance parameters must be identified. Progress towards more analog features for detection of process parameter shifts and performance degradations is also beneficial because detection and tuning in the analog domain is often faster than the digital counterpart. Hence, start-up routines could be improved with added analog tuning features. One tool to do so is the analog mismatch reduction scheme in Chap. 6. Current trends show that the conglomerate of analog and digital techniques is crucial for effective built-in tests of complex single-chip systems, motivating the continued development of BITs and digitally controllable analog circuit blocks. Pros and cons of the aforementioned self-test and calibration concepts are recapped in Table 2.2.

2.2.6 High-Volume Manufacturing Testing

A production test strategy for transceiver systems-on-a-chip has recently been proposed in [34] to address cost savings through the use of soft specification limits based on statistical parameter distributions in combination with a defect-oriented test approach that enables low-cost testing using less accurate equipment or built-in circuitry. Such a test strategy would open doors for positive impact of the circuit-level adjustment features from this research on product yields. Since the suggested approach in [34] involves crude and fast tests around the acceptable minimum and maximum specification limits for a given parameter, digital programmability in the analog blocks makes retesting with fast on-chip performance tuning possible. Therefore, in reference to Fig. 2.1, self-calibration leads to narrower parameter distributions and thus higher production yields [34].

The on-chip temperature sensor in Chap. 5 extracts the gain and linearity information that conventional power detectors [17–20] for built-in testing provide. Since such on-chip sensors generate DC output voltages, they simplify production testing by avoiding RF outputs requiring well-designed impedance-matched interfaces with the automatic test equipment (ATE). Furthermore, RF measurements drive up the production test cost and are undesirable in multi-site (parallel) testing setups due to the limited number of RF channels on the ATE [35]. Since reading out DC voltages with on-chip multiplexers is more practical than routing high-frequency signals, built-in test and calibration typically reduces the number of I/O pads, thereby decreasing die sizes.

2.2.7 Analog Tuning "Knobs"

Individual blocks are tuned as part of the system-level calibrations summarized in this chapter, for which diverse mechanisms can be used depending on the specific

Table 2.2 Comparison of transceiver built-in testing and calibration techniques

Approach	Typical applications	Advantages	Disadvantages
Digital correction and calibration (Sect. 2.2.1)	I/Q mismatch calibration	High accuracy	Large variations in the analog front-end gain or linearity cannot be corrected (e.g. saturation of analog stages from DC offset amplification)
	Digital dynamic offset compensation	No measurement circuitry in the analog front-end that could load the signal path	
	System-level performance measurements (BER, FFT, EVM) with external test input or training symbols during normal operation	Well-suited for background calibration	Convergence times are longer (millisecond range). Converge times increases with PVT variation severity
		Digital BIT circuit performance is robust to PVT variations	Adaptive optimization of analog circuits is not possible because failure cause information is not available
		Low area and power overhead (when the DSP is on the chip)	
Analog measurements and tuning (Sect. 2.2.2)	I/Q mismatch calibration in image-reject receivers	Direct correction of analog blocks with control voltages	Increased power and die area due to analog BIT circuitry
	Block-level characterization and tuning	Fast settling times	BIT circuitry is connected to CUTs and failures can impact the main signal path
	Dedicated transceiver front-end chips without on-chip digital resources	Typically suitable for background calibration	Intensive design efforts (BIT circuitry implementation is significantly different, depending on transceiver types, applications, and accuracy requirements.)
		The only option when the digital baseband processor is on a different chip	
		Can be applied to high-frequency blocks	
Loopback Testing (Sect. 2.2.3)	Production testing	The most important system-level parameter is verified: bit error rate performance	No or limited data about fault locations unless combined with analog measurement circuits
	Quick self-tests when the transceiver is idle	Fast verification of all on-chip blocks	Not suitable for on-line calibration (transceiver must be idle and in test mode)
		Low area and power overhead for BIT circuits	

(continued)

Table 2.2 (continued)

Approach	Typical applications	Advantages	Disadvantages
Combined Digital Performance Monitoring and Analog Compensation (Sect. 2.2.4)	I/Q mismatch calibration Analog dynamic offset compensation to prevent saturation Coarse start-up calibrations Production testing and on-line calibration	Analog compensation overcomes large PVT variations and reduces design margin requirements Front-end circuitry adjustments for deficiencies that cannot be corrected in the digital domain (transistors in unacceptable operating region due to process variations, low SNR from diminished front-end gain, amplified DC offsets in analog circuits that saturate internal nodes or the ADC input) Well-suited for background calibration	Limited insights into block-level performance Complex calibration algorithms Solutions are developed specific to the transceiver under test Analog circuits must be programmable
Combined Digital Monitoring, Analog Measurements, and Analog Compensation (Sect. 2.2.5)		Highest detection capability of faults and performance shifts on the block-level and system-level Block-level optimization as part of system calibration algorithms Well-suited for background calibration	Area and power overhead for measurement circuitry Complex calibration algorithms Intensive design efforts (BIT circuitry implementation is significantly different depending on transceiver types, applications, and accuracy requirements.)

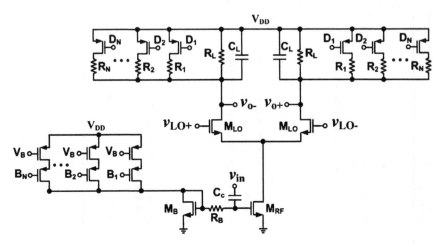

Fig. 2.9 Tuning of mixer gain ($B_1\ldots B_N$) and 2nd-order nonlinearity ($D_1\ldots D_N$)

circuit and its application. For instance, the gain of the RF transconductor in [30] has 5-bit digital gain programmability by selectively activating a number of transconductance elements that are connected in parallel. Alternatively, the transconductance values of the baseband filter in [33] are tuned by adjusting bias voltages with 8-bit DACs. Additionally, the receiver path in [33] contains 8-bit current-steering DACs to cancel DC offsets at the output of the mixing stage. Digital correction of I/Q gain mismatches can also be carried out immediately after the down-conversion by generating the bias currents for the mixers in the I and Q paths with separate current sources consisting of multiple elements [36]. This is visualized for a single-balanced mixer in Fig. 2.9, where control bits B_1–B_0 set the conversion gain. Second-order nonlinearities due to mismatches in the mixer can be reduced as well with load resistors that are comprised of multiple parts and switches [36], which enables mismatch compensation by setting the optimum resistor value for each branch at the mixer output with digital control bits D_1–D_N. Digitally programmable resistors have also been employed for enhancement of third-order nonlinearities in transconductance-capacitor baseband filters, provided that a linearization scheme with dependence on resistors is applied such as the one proposed in [37].

Circuit-level tuning methods have also been reported to recover from process variations of passive components that influence the frequency response in the RF front-end. For instance, Fig. 2.10 shows how, as proposed in [21], the input impedance matching network of a conventional inductively degenerated common-source LNA can be digitally tuned by designing it with a gate inductor L_g that is tapped at several points by closing one of the switches $S_1\ldots S_N$ to optimize the input impedance matching. However, the on-resistance of the switch in the signal path must be carefully considered during the design in order to minimize its effect on the quality factor of the input matching network as well as on the noise and linearity performance. An additional tuning feature is the varactor C_{var} in the

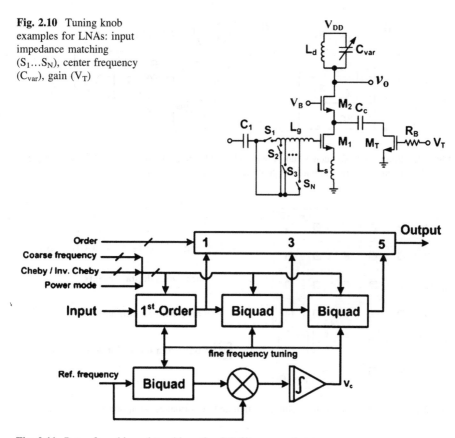

Fig. 2.10 Tuning knob examples for LNAs: input impedance matching $(S_1...S_N)$, center frequency (C_{var}), gain (V_T)

Fig. 2.11 Reconfigurable and tunable active-RC filter example

inductor–capacitor tank, which can be used to adjust the self-resonant frequency according to [38]. Finally, Fig. 2.10 also displays the gain adjustment method from [39]: the auxiliary transistor M_T is employed as variable resistor that diverts signal current to the AC ground instead of the output, modifying the LNA gain while the LNA DC bias remains unaffected thanks to the capacitor C_c.

Generally, baseband circuits allow for more tuning and reconfiguration compared to RF circuits because of the loading effects from parasitic capacitances have less impact at lower frequencies and more switches can be included in the signal path. For example, Fig. 2.11 shows the block diagram of the reconfigurable active-RC filter presented in [40], which can realize Chebyshev and Inverse Chebyshev filter functions with orders ranging from 1 to 5. Such reconfigurability enables design reuse as well as adjustable power consumption (3–7.5 mW in the discussed example) according to the filter requirements. Moreover, the design in [40] permits filter cutoff frequency tuning by two means that are displayed in Fig. 2.12: coarse tuning with digitally-controlled capacitors (switches S_0–S_2), and continuous fine tuning with an analog control voltage (V_C).

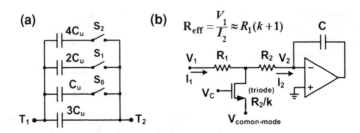

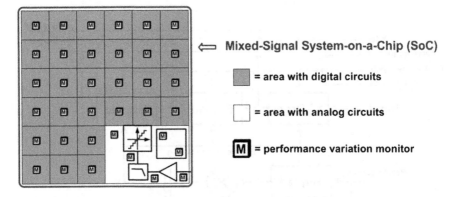

Fig. 2.12 Filter cutoff frequency tuning with adjustable elements: **a** coarse tuning with programmable capacitors, **b** fine tuning with continuous impedance multipliers

Fig. 2.13 Partitioned die with on-chip variation and performance monitoring

2.2.8 Variation-Aware Design of Digital Circuits

The purpose of this subsection is to distinguish the variation-aware design approaches for entirely digital SoCs from those for mixed-signal SoCs. Both analog and digital variation-aware design approaches require on-chip PVT monitors or measurement circuits. As visualized in Fig. 2.13, a die can be divided into many partitions to detect within-die variation, where each section contains at least one local monitoring circuit. For digital systems, variation monitors have been reported with features such as ring oscillators or delay lines for speed assessments [41–44] and temperature sensors for power density management [45–48]. As the levels of integration and number of processor cores increase (e.g. 80 cores in [49]), the adaptive methods will become more effective when the number of partitions with local PVT monitors is also increased. Nevertheless, the die area of the monitors and routing must be minimized to avoid excessive fabrication cost.

In microprocessors and other digitally-intensive systems it becomes increasingly popular to manage on-chip power dissipations and temperatures using numerous variable supply voltages or clock frequencies for different sections (cores) on the die, such as in [50–52]. These techniques directly benefit from the information provided by the distributed placement of the sensors with sensitivity to static and dynamic power.

A major advantage of variation-sensing approaches for on-chip calibration of circuits is the enhanced resilience to the process and environmental variations that are presently creating yield and reliability challenges for chips fabricated with widely used CMOS technology. Since the threshold voltage is a significant process variation indicator for analog [53, 54] and digital circuits [41, 55], there are existing methods to monitor its statistical variation [8, 56]. In digital sections, the local operating frequency/speed measurements supplied by the variation monitors is also valuable information in adaptive body bias methods and other approaches to cope with worsening within-die variations in CMOS technologies [57–60]. In digitally-intensive systems, the extracted information that represents local on-die variations is sufficient to enable on-chip power and thermal management techniques by applying variable supply voltages or clock frequencies in the different sections (cores) [50–52, 61]. In general, the continued enhancement of on-chip local variation-sensing capabilities to assess the digital performance indicators will allow more reductions of variation and aging effects [45]. To achieve variation-resilient analog circuits in mixed-signal SoCs, the variation monitors are typically placed directly next to analog blocks as indicated in Fig. 2.13 because the information about local parameter variations is not sufficient to predict performance. As elaborated throughout this book, the need to directly extract critical performance indicators for individual analog blocks can be addressed with dedicated sensors. This difference generally leads to more specialized and complex measurement and calibration procedures compared to the digital counterparts.

References

1. C. Chiang, J. Kawa, *Design for Manufacturability and Yield for Nano-scale CMOS* (Springer, Dordrecht, 2007), pp. 14–15
2. W. Zhao, Y. Cao, F. Liu, K. Agarwal, D. Acharyya, S. Nassif, K. Nowka, Rigorous extraction of process variations for 65 nm CMOS design, in *Proceedings of European Solid-State Device Research Conference (ESSDERC)*, Sept 2007, pp. 89–92
3. G.W. Roberts, B. Dufort, Making complex mixed-signal telecommunication integrated circuits testable. IEEE Commun. Mag. 90–96 (1999)
4. A. Zjajo, J.P. de Gyvez, Evaluation of signature-based testing of RF/analog circuits. Proc. Eur.Test Symp. 62–67 (2005)
5. G.G.E. Gielen, Design methodologies and tools for circuit design in CMOS nanometer technologies, in *Proceedings of European Solid-State Device Research Conference (ESSDERC)*, Sept 2006, pp. 21–32
6. H. Masuda, M. Tsunozaki, T. Tsutsui, H. Nunogami, A. Uchida, K. Tsunokuni, A novel wafer-yield PDF model and verification with 90–180 nm SOC chips. IEEE Trans. Semicond. Manuf. 21(4), 585–591 (2008)
7. V.A. Zivkovic, F. van der Heyden, G. Gronthoud, F. de Jong, Analog test bus infrastructure for RF/AMS modules in core-based design, in *Proceedings of 13th European Test Symposium*, May 2008, pp. 27–32

8. K. Agarwal, J. Hayes, S. Nassif, Fast characterization of threshold voltage fluctuation in MOS devices. IEEE Trans. Semicond Manuf. **21**(4), 526–533 (2008)
9. J.P.F. Glas, Digital I/Q imbalance compensation in a low-IF receiver, in *Proceedings of IEEE Global Telecommunications Conference (GLOBECOM)*, vol. 3, Nov 1998, pp. 1461–1466
10. W. Eberle, J. Tubbax, B. Come, S. Donnay, H. De Man, G. Gielen, OFDM-WLAN receiver performance improvement using digital compensation techniques, in *Proceedings IEEE Radio and Wireless Conference (RAWCON)*, Aug 2002, pp. 111–114
11. I. Elahi, K. Muhammad, P.T. Balsara, I/Q mismatch compensation using adaptive decorrelation in a low-IF receiver in 90 nm CMOS process. IEEE J Solid-State Circuits **41**(2), 395–404 (2006)
12. B. Shi, Y. W. Chia, An analog mismatch calibration system for image-reject receivers, in *Proceedings of European Conference on Wireless Technology*, Oct 2005, pp. 225–228
13. R.B. Staszewski, I. Bashir, O. Eliezer, RF Built-in self test of a wireless transmitter. IEEE Trans Circuits Syst Express Briefs **54**, 186–190 (2007)
14. R. Montemayor, B. Razavi, A self-calibrating 900 MHz CMOS image-reject receiver, in *Proceedings of Euopean Solid-State Circuits Conference (ESSCIRC)*, Sept 2000, pp. 320–323
15. M.A.I. Elmala, S.H.K. Embabi, Calibration of phase and gain mismatches in weaver image-reject receiver. IEEE J Solid-State Circuits **39**(2), 283–289 (2004)
16. J.-Y. Ryu, B.C. Kim, I. Sylla, A new low-cost RF built-in self-test measurement for system-on-chip transceivers. IEEE Trans. Instrum. Meas. **55**(2), 381–388 (2006)
17. Q. Yin, W.R. Eisenstadt, R.M. Fox, T. Zhang, A translinear RMS detector for embedded test of RF ICs. IEEE Trans. Instrum. Meas. **54**(5), 1708–1714 (2005)
18. S. Bhattacharya, A. Chatterjee, Use of embedded sensors for built-in-test RF circuits, in *Proceedings of IEEE International Test Conference (ITC)*, Oct 2004, pp. 801–809
19. Q. Wang, M. Soma, RF front-end system gain and linearity built-in test, in *Proceedings of 24th IEEE VLSI Test Symposium*, May 2006, pp. 228–233
20. A. Valdes-Garcia, R. Venkatasubramanian, J. Silva-Martinez, E. Sánchez-Sinencio, A broadband CMOS amplitude detector for on-chip RF measurements. IEEE Trans. Instrum. Meas. **57**(7), 1470–1477 (2008)
21. T. Das, A. Gopalan, C. Washburn, P.R. Mukund, Self-calibration of input-match in RF front-end circuitry. IEEE Trans. Circuits Syst. Express Briefs **52**(12), 821–825 (2005)
22. V. Stopjakova, H. Manhaeve, M. Sidiropulos, On-chip transient current monitor for testing of low-voltage CMOS IC, in *Proceedings of Design, Automation and Test in Europe Conference and Exhibition*, Mar 1999, pp. 538–542
23. A.P. Jose, K.A. Jenkins, S.K. Reynolds, On-chip spectrum analyzer for analog built-in self test, in *Proceedings of IEEE VLSI Test Symposium*, May 2005, pp. 131–136
24. A. Valdes-Garcia, F.A.-L. Hussien, J. Silva-Martinez, E. Sánchez-Sinencio, An integrated frequency response characterization system with a digital interface for analog testing. IEEE J Solid-State Circuits **41**(10), 2301–2313 (2006)
25. J.J. Dabrowski, R.M. Ramzan, Built-in loopback test for IC RF transceivers. IEEE Trans. Very Large Scale Integr. VLSI Syst. **18**(6), 933–946 (2010)
26. M. Onabajo, J. Silva-Martinez, F. Fernandez, E. Sánchez-Sinencio, An on-chip loopback block for RF transceiver built-in test. IEEE Trans. Circuits Syst. Express Briefs **56**(6), 444–448 (2009)
27. G. Srinivasan, A. Chatterjee, F. Taenzler, Alternate loop-back diagnostic tests for wafer-level diagnosis of modern wireless transceivers using spectral signatures, in *Proceedings 24th VLSI Test Symposium*, May 2006, pp. 222–227
28. A. Haider, S. Bhattacharya, G. Srinivasan, A. Chatterjee, A system-level alternate test approach for specification test of RF transceivers in loopback mode, in *Proceedings of 18th International Conference on VLSI Design*, Jan 2005, pp. 289–294
29. M. Negreiros, L. Carro, A.A. Susin, An improved RF loopback for test time reduction, in *Proceedings of Design, Automation, and Test in Europe Conference and Exhibition*, Mar 2006, pp. 646–651

30. D. Kaczman, M. Shah, M. Alam, M. Rachedine, D. Cashen, L. Han, A. Raghavan, A single-chip 10-band WCDMA/HSDPA 4-band GSM/EDGE SAW-less CMOS receiver with DigRF 3G interface and +90 dBm IIP2. IEEE J. Solid-State Circuits **44**(3), 718–739 (2009)

31. H. Darabi, J. Chiu, S. Khorram, H.J. Kim, Z. Zhou, H.-M. Chien, B. Ibrahim, E. Geronaga, L.H. Tran, A. Rofougaran, A dual-mode 802.11b/Bluetooth radio in 0.35 μm CMOS. IEEE J. Solid-State Circuits **40**(3), 698–706 (2005)

32. Y.-H. Hsieh, W.-Y. Hu, S.-M. Lin, C.-L. Chen, W.-K. Li, S.-J. Chen, D.J. Chen, An auto-I/Q calibrated CMOS transceiver for 802.11g. IEEE J. Solid-State Circuits **40**(11), 2187–2192 (2005)

33. I. Vassiliou, K. Vavelidis, T. Georgantas, S. Plevridis, N. Haralabidis, G. Kamoulakos, C. Kapnistis, S. Kavadias, Y. Kokolakis, P. Merakos, J.C. Rudell, A. Yamanaka, S. Bouras, I. Bouras, A single-chip digitally calibrated 5.15–5.825 GHz 0.18 μm CMOS transceiver for 802.11a wireless LAN. IEEE J. Solid-State Circuits **38**(12), 2221–2231 (2003)

34. O. Eliezer, R.B. Staszewski, D. Mannath, A statistical approach for design and testing of analog circuitry in low-cost SoCs, in *Proceedings of IEEE International Midwest Symposium on Circuits and Systems (MWSCAS)*, Aug 2010, pp. 461–464

35. M. Onabajo, F. Fernandez, J. Silva-Martinez, E. Sánchez-Sinencio, Strategic test cost reduction with on-chip measurement circuitry for RF transceiver front-ends—an overview. Proc. IEEE Int. Midwest Symp. Circuits Syst. (MWSCAS) **2**, 643–647 (2006)

36. K. Kivekas, A. Parssinen, J. Ryynanen, J. Jussila, K. Halonen, Calibration techniques of active BiCMOS mixers. IEEE J. Solid-State Circuits **37**(6), 766–769 (2002)

37. M. Mobarak, M. Onabajo, J. Silva-Martinez, E. Sánchez-Sinencio, Attenuation- predistortion linearization of CMOS OTAs with digital correction of process variations in OTA-C filter applications. IEEE J. Solid-State Circuits **45**(2), 351–367 (2010)

38. N. Ahsan, J. Dabrowski, A. Ouacha, A self-tuning technique for optimization of dual band LNA, in *Proceedings of Euopean Conference Wireless Technology (EuWiT)*, Oct 2008, pp. 178–181

39. C.-H. Liao, H.-R. Chuang, A 5.7 GHz 0.18 μm CMOS gain-controlled differential LNA with current reuse for WLAN receiver. IEEE Microwave Compon. Lett. **13**(12), 526–528 (2003)

40. H. Amir-Aslanzadeh, E.J. Pankratz, E. Sánchez-Sinencio, A 1 V +31 dBm IIP3, reconfigurable, continuously tunable, power-adjustable active-RC LPF. IEEE J. Solid-State Circuits **44**(2), 495–508 (2009)

41. M. Miyazaki, G. Ono, K. Ishibashi, A 1.2-GIPS/W microprocessor using speed-adaptive threshold-voltage CMOS with forward bias. IEEE J. Solid-State Circuits **37**(2), 210–217 (2002)

42. K.A. Bowman, J.W. Tschanz, S.L. Lu, P.A. Aseron, M.M. Khellah, A. Raychowdhury, B.M. Geuskens, C. Tokunaga, C.B. Wilkerson, T. Karnik, V.K. De, A 45 nm resilient microprocessor core for dynamic variation tolerance. IEEE J. Solid-State Circuits **46**(1), 194–208 (2011)

43. Y.-B. Kim, K. K. Kim, J. Doyle, A CMOS low power fully digital adaptive power delivery system based on finite state machine control, in *Proceedings of IEEE International Symposium Circuits and Systems (ISCAS)*, May 2007, pp. 1149–1152

44. M. Bhushan, A. Gattiker, M.B. Ketchen, K.K. Das, Ring oscillators for CMOS process tuning and variability control. IEEE Trans. Semicond. Manuf. **19**(1), 10–18 (2006)

45. J. Tschanz, N. S. Kim, S. Dighe, J. Howard, G. Ruhl, S. Vanga, S. Narendra, Y. Hoskote, H. Wilson, C. Lam, M. Shuman, C. Tokunaga, D. Somasekhar, S. Tang, D. Finan, T. Karnik, N. Borkar, N. Kurd, V. De, Adaptive frequency and biasing techniques for tolerance to dynamic temperature-voltage variations and aging, in *IEEE International Solid-State Circuits Conference (ISSCC) Digest of Technical Papers*, Feb 2007, pp. 292–604

46. S.-C. Lin, K. Banerjee, A design-specific and thermally-aware methodology for trading-off power and performance in leakage-dominant CMOS technologies. IEEE Trans. Very Large Scale Integr. VLSI Syst. **16**(11), 1488–1498 (2008)

47. K. Woo, S. Meninger, T. Xanthopoulos, E. Crain, D. Ha, D. Ham;, Dual-DLL-based CMOS all-digital temperature sensor for microprocessor thermal monitoring, in *IEEE International*

Solid-State Circuits Conference (ISSCC) Digest of Technical Papers, Feb 2009, pp. 68–69, 69a

48. P. Ituero, J.L. Ayala, M. Lopez-Vallejo, A nanowatt smart temperature sensor for dynamic thermal management. IEEE Sens. J. 8(12), 2036–2043 (2008)

49. S. Dighe, S.R. Vangal, P. Aseron, S. Kumar, T. Jacob, K.A. Bowman, J. Howard, J. Tschanz, V. Erraguntla, N. Borkar, V.K. De, S. Borkar, Within-die variation-aware dynamic-voltage-frequency-scaling with optimal core allocation and thread hopping for the 80-core TeraFLOPS processor. IEEE J. Solid-State Circuits 46(1), 184–193 (2011)

50. T. Fischer, J. Desai, B. Doyle, S. Naffziger, B. Patella, A 90 nm variable frequency clock system for a power-managed itanium architecture processor. IEEE J. Solid-State Circuits 41(1), 218–228 (2006)

51. N. Drego, A. Chandrakasan, D. Boning, D. Shah, Reduction of variation-induced energy overhead in multi-core processors. IEEE Trans. Comput. Aided Des. Integr. Circuits Syst. 30(6), 891–904 (2011)

52. A. Allen, J. Desai, F. Verdico, F. Anderson, D. Mulvihill, D. Krueger, Dynamic frequency-switching clock system on a quad-core Itanium® processor, in IEEE International Solid-State Circuits Conference (ISSCC) Digest of Technical Papers, Feb 2009, pp. 62–63, 63a

53. K. Lakshmikumar, R. Hadaway, M.A. Copeland, Characterization and modeling of mismatch in MOS transistors for precision analog design. J. Solid-State Circuits 21(12), 1057–1066 (1986)

54. P.R. Kinget, Device mismatch and tradeoffs in the design of analog circuits. J. Solid-State Circuits 40(6), 1212–1224 (2005)

55. K.K. Kim, W. Wang, K. Choi, On-chip aging sensor circuits for reliable nanometer MOSFET digital circuits. IEEE Trans. Circuits Syst. Express Briefs 57(10), 798–802 (2010)

56. R. Rao, K.A. Jenkins, J.-J. Kim, A local random variability detector with complete digital on-chip measurement circuitry. IEEE J. Solid-State Circuits 44(9), 2616–2623 (2009)

57. N. Mehta, B. Amrutur, Dynamic supply and threshold voltage scaling for CMOS digital circuits using in situ power monitor, to appear in IEEE Transaction on Very Large Scale Integration (VLSI) Systems

58. K.K. Kim, Y.-B. Kim, A novel adaptive design methodology for minimum leakage power considering PVT variations on nanoscale VLSI systems. IEEE Trans. Very Large Scale Integr. VLSI Syst. 17(4), 517–528 (2009)

59. J.W. Tschanz, J.T. Kao, S.G. Narendra, R. Nair, D.A. Antoniadis, A.P. Chandrakasan, V. De, Adaptive body bias for reducing impacts of die-to-die and within-die parameter variations on microprocessor frequency and leakage. IEEE J. Solid-State Circuits 37(11), 1396–1402 (2002)

60. M. Mostafa, M. Anis, M. Elmasry, On-chip process variations compensation using an analog adaptive body bias (A-ABB), to appear in IEEE Transactions on Very Large Scale Integration (VLSI) Systems

61. R. McGowen, C.A. Poirier, C. Bostak, J. Ignowski, M. Millican, W.H. Parks, S. Naffziger, Power and temperature control on a 90 nm itanium family processor. IEEE J. Solid-State Circuits 41(1), 229–237 (2006)

Chapter 3
High-Linearity Transconductance Amplifiers with Digital Correction Capability

Abstract The push towards wider bandwidths in baseband filter applications calls for operational transconductance amplifiers (OTAs) with progressively better linearity at higher frequencies. In this chapter, an architectural solution is described that can be applied to diverse circuit-level OTA configurations. Effective linearization over a wide frequency range demands a mechanism to correct for high-frequency effects and process variations. Accordingly, digital programmability to ensure high linearity and compatibility with modern CMOS technologies is discussed. The linearization technique utilizes two matched OTAs to cancel harmonic distortion components, creating a robust architecture. Compensation for process variations and frequency-dependent distortion based on Volterra series analysis is achieved by employing a delay equalization scheme with on-chip programmable resistors.

3.1 Background

Operational transconductance amplifiers (OTAs) are essential elements of trans-conductance-capacitor (G_m-C) filters [1, 2], $\Delta\Sigma$ modulators [3], gyrators, variable-gain amplifiers, and negative-resistance elements. Compared to their active-RC counterparts, G_m-C filters enable low-power operation and tuning of the filter characteristics at higher frequencies, but are less linear. Tunable active-RC filters are suitable for low-frequency applications; however, extending their use to higher

This chapter includes portions reprinted with permission, from "Attenuation-predistortion linearization of CMOS OTAs with digital correction of process variations in OTA-C filter applications," M. Mobarak, M. Onabajo, J. Silva-Martinez, and E. Sánchez-Sinencio, *IEEE J. Solid-State Circuits*, vol. 45, no. 2, pp. 351–367, Feb. 2010, © 2010 IEEE.

frequencies would require significantly more power. On the other hand, OTA-based filters in wireless receivers and continuous-time (CT) $\Delta\Sigma$ analog-to-digital converters (ADCs) increasingly mandate better linearity at higher frequencies. These applications typically require highly linear OTAs with third-order intermodulation (IM3) distortion better than -60 dB. Further advances in high-frequency G_m-C filters with SNDRs over 50 dB are also desirable for channel selection/equalization in multi-Gbps portable data communication devices [2], and for possible application in next generation analog-to-information receivers if dynamic ranges greater than 90 dB in 200 MHz bandwidth become attainable [4].

Viable high-frequency G_m-C filter solutions were presented in [1, 5] with 3 dB frequencies at 275 and 184 MHz, respectively. The topology reported in [1] has low noise, limited linearity, and a pseudo-differential realization prone to low power supply rejection ratio (PSRR). The filter in [5] achieves high linearity with relatively low power but higher noise. Trade-offs between linearity, noise, power, and operating frequency are common, which have been incorporated into figures of merit (FOMs) such as in [6, 7]. Recent works also address alternative filter structures such as the source-follower-based approach [8] and performance improvements of typical OTA topologies [9].

A popular linearization approach is to cross-couple two transconductors, theoretically canceling certain harmonics at specific bias conditions over a limited frequency range. A typical cross-coupled OTA contains two paths; each having different transconductance and ideally the same amount of harmonic distortion. When cross-coupled, the equal harmonics cancel under ideal conditions and the effective transconductance is the difference between the two paths. The frequency dependence of this approach has been analyzed with a Volterra series in [10, 11], in which the analytical expressions are correlated with measurement results. Process-voltage–temperature (PVT) variations, high-frequency effects, and device modeling inaccuracies will create unforeseen mismatches between the two amplifiers. Therefore, precision tuning of bias currents/voltages is typically required. Attenuation and cross-coupling have been combined for the low-noise amplifier in [12], in which distortion cancellation is restricted to third-order nonlinearities with a feed forward path and precise off-chip input attenuation.

The methodology to be discussed in this chapter was proposed in [13]. It is an architectural solution that achieves up to 22 dB IM3 improvement over an identical nonlinearized OTA design at frequencies as high as 350 MHz. It can be generalized to fully-differential topologies which offer high PSRR and common-mode rejection ratio (CMRR). Since the maximum frequency is mainly limited by process parasitics and OTA performance, the approach shows promise of exceeding 350 MHz bandwidth in future nanoscale CMOS processes. Robust linearization over a wide frequency range demands a mechanism to correct for high-frequency effects and PVT variations, for which a digital programmability scheme is included. Section 3.2 describes the attenuation-predistortion linearization methodology along with the result from Volterra analysis that furnishes a design criterion to ensure broadband performance. The corresponding OTA and G_m-C filter design issues are addressed in Sect. 3.3. Section 3.4 presents digital correction requirements based on PVT simulations. Measurement results for a linearized fully-differential OTA and a second-order biquadratic G_m-C lowpass filter in 0.13 μm CMOS technology are provided in Sect. 3.5, and conclusions are given in Sect. 3.6.

3.2 Attenuation-Predistortion Linearization Methodology

Signal attenuation at the OTA input [10] reduces the effective transconductance and decreases the SNR. Alternatively, distortion cancellation by means of cross-coupled differential pairs results in increased power consumption and noise level proportional to the transistor parameters in the additional path. Since the extra differential pair normally has less transconductance than the main pair, the effective transconductance is reduced by 10–50%. However, both transistor pairs should have the same third-order nonlinearity, which translates into different transistor sizes and bias currents for each pair. As a result, the cross-coupling technique is sensitive to PVT variations and restricted to narrow frequency ranges. Another common method to linearize a transistor having transconductance g_m is to add a degeneration resistor R_{sd} at the source [10], which makes the third-order harmonic distortion proportional to the factor $1/(1 + g_m R_{sd})^3$. Nonetheless, large degeneration resistance results in higher input-referred noise, lower transconductance, and less voltage headroom. The effective transconductance (g_{msd}) and the input-referred noise (v_{nsd}^2) with *resistive source degeneration* are given by

$$g_{msd} = \frac{g_m}{1 + g_m R_{sd}}, \quad v_{nsd}^2 \approx \frac{4kT}{g_m} (2/3 + g_m R_{sd}) \tag{3.1}$$

where the noise coefficient γ was approximated as 2/3 assuming long-channel devices (but this factor could be even >1 for short-channel transistors). For example, using a degeneration factor $g_m R_{sd} = 2$ will ideally result in IM3 improvement of approximately 29 dB, an input-referred noise power increase by a factor of 4, and a decrease of the transconductance to one third of its original value. But based on simulations of the OTA under investigation in this chapter with $g_m R_{sd} = 2$, the expected IM3 improvement would be 25.2 dB with an associated noise power increase of more than 9 times.

The attenuation-predistortion method is independent of OTA topology and involves cancellation of all distortion components except those from secondary effects at high frequencies. It can be combined with other circuit-level linearization techniques internal to the OTA, such as source degeneration or cross-coupling.

3.2.1 Single-Ended Circuits

Figure 3.1 depicts the single-ended architecture that contains an auxiliary branch with an OTA having identical dimensions, DC bias, and AC common-mode

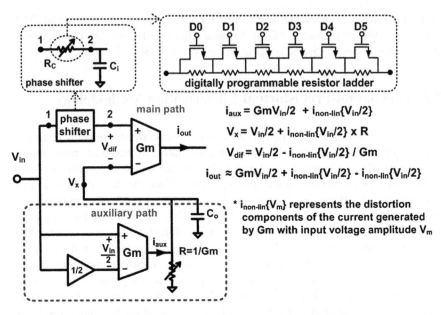

$$i_{aux} = Gm V_{in}/2 + i_{non\text{-}lin}\{V_{in}/2\}$$

$$V_x = V_{in}/2 + i_{non\text{-}lin}\{V_{in}/2\} \times R$$

$$V_{dif} = V_{in}/2 - i_{non\text{-}lin}\{V_{in}/2\} / Gm$$

$$i_{out} \approx Gm V_{in}/2 + i_{non\text{-}lin}\{V_{in}/2\} - i_{non\text{-}lin}\{V_{in}/2\}$$

* $i_{non\text{-}lin}\{V_m\}$ **represents the distortion components of the current generated by Gm with input voltage amplitude V_m**

Fig. 3.1 Attenuation-predistortion linearization for single-ended circuits

conditions as in the main path to generate the distortion components required for cancellation. An important advantage of identical paths is robustness to PVT variations because of optimal device matching obtainable from proper layout. In this scheme, it is avoided to base the distortion cancellation on branches with different transconductor device dimensions or bias conditions, which would degrade matching accuracy. But even with minimized mismatches, nonlinearities are particularly frequency-dependent at high frequencies and remain sensitive to PVT variations as established in Sect. 3.4. Hence, the linearization method involves variable resistors to tune performance and counteract high-frequency degradation as well as PVT variations. Either a resistive or capacitive divider can form the attenuator at the input of the auxiliary path; however, resistors would add more noise.

Distortion cancellation in the single-ended case requires $G_m \times R = 1$, which is ascertained by the following analysis. For a certain input voltage amplitude V_m, the output current can be divided into a linear part $i_{lin}\{V_m\} = G_m \times V_m$ and a nonlinear part $i_{non\text{-}lin}\{V_m\} = g_{m2} \times V_m^2 + g_{m3} \times V_m^3 + ...$, where $g_{m2}, g_{m3},...$ are Taylor series coefficients of the transconductance. The differential input of the main OTA is: $V_{dif} = V_{in} - [V_{in}/2 + i_{non\text{-}lin}\{V_{in}/2\}/G_m] = V_{in}/2 - i_{non\text{-}lin}\{V_{in}/2\}/G_m$. Under ideal conditions, the distortion generated in the auxiliary path, $-i_{non\text{-}lin}\{V_{in}/2\}$, cancels out the distortion in the main voltage-to-current conversion. In practice, distortion components caused by nonlinearities at the output of the auxiliary OTA and high-frequency effects introduce some finite uncancelled distortion. Capacitor C_o

represents the lumped output capacitance of the auxiliary OTA, input capacitance of the main OTA, and layout parasitics. Resistor R_c of the phase shifter and equivalent input capacitance C_i provide first-order frequency compensation, creating a pole to equalize the phase shift between the main and auxiliary paths. Compensation is necessary at high frequencies because C_o at the negative input terminal of the main OTA creates a pole with resistor R in the auxiliary path.

3.2.2 Fully-Differential Circuits

A conceptual diagram of the attenuation-predistortion linearization approach for a fully-differential transconductor (G_m) is displayed in Fig. 3.2. In the fully-differential case, attenuation factors at the input of the transconductors are realized with floating-gate devices described in Sect. 3.3.1. As discussed in [10, 14], the inherent input attenuation with floating-gate stages enhances the OTA linearity. The distortion cancellation principle is the same as in the single-ended case, but different conditions must be satisfied for fully-differential implementation, which are explained in Sects. 3.2.3 and 3.3.1 with regards to the attenuation ratios. By selecting an input attenuation ratio of 1/3 and voltage gain of 3 in the auxiliary branch $(G_m \times R = 3)$, the signal amplitude V_x is equal to V_{in} plus three times the distortion components caused by the nonlinear current $i_{non-lin}\{V_{in}/3\}$ from the transconductor with input amplitude of $V_{in}/3$. In the main path, the effective differential OTA input signal is: $V_{dif} = 2\ V_{in}/3 - V_x/3 = 2V_{in}/3 - [V_{in} + 3 \times i_{non-lin}\{V_{in}/3\}/G_m]/3 = V_{in}/3 - i_{non-lin}\{V_{in}/3\}/G_m$. Thus, the differential signal contains the attenuated input signal and the inverse of the distortion generated by the identical G_m in the auxiliary branch for distortion cancellation during the voltage-to-current conversion in the main path. Ideally, the distortion components are canceled by the equal and opposite terms from the predistortion of the differential input signal except for negligible higher-order components.

C_o in Fig. 3.2 represents the equivalent differential capacitance of all parasitic capacitances at the output of the auxiliary OTA, and C_p is the differential equivalent of the parasitic capacitances at the input of the main OTA. Expressions for optimum distortion cancellation at high frequencies are provided in Sect. 3.2.4. Linear RC phase shifter networks are chosen for the distortion cancellation and frequency compensation implementation. Resistors R and R_c are tuned with 6-bit resolution to compensate for mismatches/PVT variations. The phase shifter block is utilized to equalize the delay from the input to summing nodes 3 and 4 in Fig. 3.2. Furthermore, the phase shifter enables optimization of the nonlinearity cancellation based on high-frequency effects.

3.2.3 Scaling of Attenuation Ratios

Depending on application-specific requirements, the design parameters in the attenuation-predistortion linearization approach can be selected to adjust the

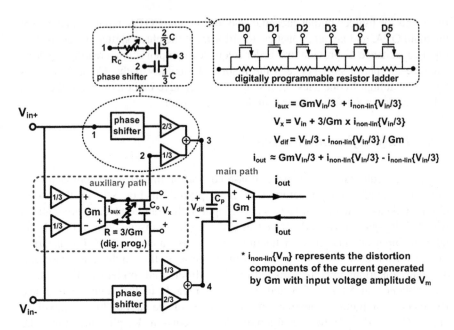

Fig. 3.2 Attenuation-predistortion linearization for fully-differential circuits

voltage swings and the effective transconductance. Figure 3.3 shows the differential attenuation-predistortion linearization scheme, where frequency compensation and parasitic capacitors have been omitted for simplicity. The following analysis assumes floating gates as a practical attenuator implementation choice under the constraint that factors k_1 and $(1 - k_1)$ are related as elaborated upon in Sect. 3.3.1, but less restrictive types of attenuators could also be used. The output current i_o of an OTA due to an input voltage V_m can be modeled as having a linear and a nonlinear part: $i_o = G_m V_m + i_{non-lin}\{V_m\}$. Ignoring high-frequency and secondary effects, the following relation can be written:

$$i_{out} \approx G_m[k_1 - (1 - k_1)k_2 G_m R]V_{in} - (1 - k_1)G_m R \cdot i_{non-lin}\{k_2 V_{in}\} \\ + i_{non-lin}\{[k_1 - (1 - k_1)k_2 G_m R]V_{in}\} \tag{3.2}$$

where: $i_{non-lin}\{k_2 V_{in}\} \cdot R(1 - k_1) \ll (k_1 - (1 - k_1)k_2 G_m R)V_{in}$ is assumed in the approximation.

To cancel the distortion, the following conditions should hold:

(1) The auxiliary and main OTAs should have the same effective input voltage amplitudes such that an identical distortion is created at their respective outputs.

(2) The gain in the auxiliary path must ensure that the distortion through this signal path reaches the output of the main OTA with a gain of -1.

Fig. 3.3 Low-frequency
model for the attenuation-
predistortion scheme

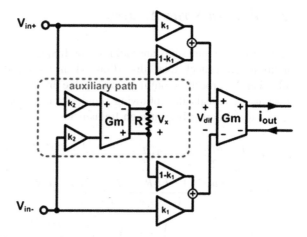

(3) The internal signal swings should be bounded, i.e.:

$$k_2 G_m R \leq 1 \tag{3.3}$$

Applying conditions (1) and (2), cancellation of the nonlinear terms in (3.2)
requires:

$$(1 - k_1)G_m R = 1, \quad k_2 = k_1/2 \tag{3.4}$$

Consequently, the effective transconductance with linearization is given by

$$G_{meff} = [k_1 - (1 - k_1)k_2 G_m R]G_m = (k_1/2)G_m = k_2 G_m \tag{3.5}$$

Condition (3) depends on the application and is not always necessary.
Cancellation of distortion with the attenuation-predistortion technique requires
weakly nonlinear operation in the auxiliary branch, which is ensured by limiting the
signal swing with this condition. The example that is presented in Fig. 3.2 was
derived with $k_2 G_m R = 1$, ensuring that the signal swing at the output of the auxiliary
OTA is the same as at its input. This choice was made to maintain the same maximum
input voltage swing as the initial OTA without saturating the OTA in the linearization
path. If the specified input signal is $k_2 G_m R$ times below the OTA saturation level, then
k_2 can be increased accordingly to obtain $k_2 G_m R > 1$ and higher effective transconductance based on (3.5). But, this choice is only permissible if a reduction of the
maximum input swing by $k_2 G_m R$ can be tolerated, which would imply a reduction in
the dynamic range. Typically, choosing $k_2 G_m R = 1$ is advantageous to maintain the
same maximum input voltage swing as the original OTA after the linearization.
Selection of $k_1 = 2/3$ and $k_2 = 1/3$ results in the highest effective transconductance
that can be achieved in (3.5) based on the above conditions while also satisfying the
attenuation factor relationships in the floating-gate devices (Sect. 3.3.1) with identical signal swings at the input and output of the auxiliary OTA ($k_2 G_m R = 1$). Hence,
$G_m R = 3$ under the stated conditions.

3.2.4 Volterra Series Analysis

The preceding expressions are valid at low frequencies and give insights into the conditions to cancel total distortion when secondary effects are negligible. Following the procedure outlined in [15], the 3rd-order Volterra series analysis in Appendix A reveals the following requirement for the phase shifter resistor in Fig. 3.2 to minimize IM3 at high frequencies:

$$
\begin{aligned}
i_{IM3} \approx\; & g_{m3}\left(\frac{k_1/2}{1+2C_p/C}\right)^3 \left(3V_{in1}^2 V_{in2}/4\right)\left(\frac{1+j\omega_1 C((1-k_1)R-k_1 R_c)+2j\omega_1 C_o R}{1+j\omega_1 b-c\omega_1^2}\right)^2 \\
& \times \left(\frac{1-j\omega_1 C((1-k_1)R-k_1 R_c)-2j\omega_1 C_o R}{1-j\omega_1 b-c\omega_1^2}\right) \\
& - g_{m3}\left(\frac{k_1/2}{1+2C_p/C}\right)^3 \left(3V_{in1}^2 V_{in2}/4\right)\frac{1+j\omega_1 C k_1 R_c}{1+j\omega_1 b-c\omega_1^2} \\
\Rightarrow\; & R_c \approx \frac{(1-k_1)+2C_o/C}{2k_1}R \quad for \;\; i_{IM3}\approx 0
\end{aligned}
\tag{3.6}
$$

In the discussed example case with $k_1 = 2/3$, the condition to cancel IM3 with the phase shifter block in Fig. 3.2 is $R_c = (R/4)\cdot(1+6C_o/C)$. To ensure high linearity with variations of parasitic capacitances, the programmable range of R_c is selected based on process corner simulations as described in Sect. 3.4.

3.3 Circuit-Level Design Considerations

3.3.1 Fully-Differential OTA with Floating-Gate Transistors

Figure 3.4 displays the schematic of the OTAs implemented on the 0.13 μm CMOS test chip with a 1.2 V supply. Attenuators k_1, $(1 - k_1)$, and k_2 are realized with floating-gate devices for attenuation-predistortion linearization of this fully-differential topology. The gates (G) of the standard NMOS transistors in the OTA core are not resistively biased and are only connected to two conventional metal-insulator-metal (MIM) capacitors. Figure 3.4 also visualizes the equivalent capacitive load seen at the V_{1+} and V_{1-} inputs, where C_{pt} represents the effective gate-to-ground(AC) capacitance from transistor parasitic capacitances. With this configuration, the gate voltages are: $V_{G+/-} = (C_{FG1}/C_{total})V_{1+/-} + (C_{FG2}/C_{total})V_{2+/-}$, where $C_{total} \approx C_{FG1} + C_{FG2}$ when C_{pt} is negligible. It follows that the attenuation factors in Fig. 3.3 are: $C_{FG1}/C_{total} = k_1$ and $C_{FG2}/C_{total} = (C_{total} - C_{FG1})/C_{total} = 1 - k_1$. The accuracy of the k_1 and $(1 - k_1)$ factors predominantly depends on the matching of the MIM capacitors C_{FG1} and C_{FG2}, which can be achieved within 0.1–1% using proper layout techniques. As assessed in

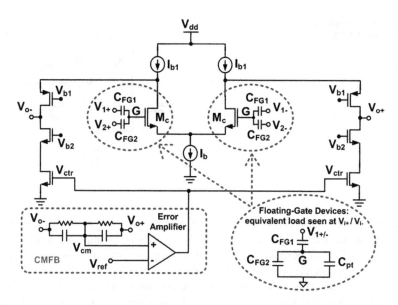

Fig. 3.4 Folded-cascode OTA (implements G_m in the main and auxiliary paths)

Sect. 3.4, such a matching accuracy is more than sufficient with the 3%-step programmability of resistor R for gain mismatch compensation in both paths.

In the layout, all nodes G at the floating gates in Fig. 3.4 are connected to the top metal layer using standard poly-metal contacts and metal–metal vias. During fabrication, this connection ensures that any charge stored on the floating gates flows to the substrate because all connections to the top metal are still joined prior to their separation during the last etching step. Thus, no charge is stored on the floating gates when the substrate contacts are also connected to the top metal layer [16], allowing gate discharge into the substrate before the last etching operation. After etching, the top metal extensions of the gates without trapped charge are floating, leaving only the connections to the two MIM capacitors. The floating-gate device design expressions for k_1 and $(1 - k_1)$ above are assuming absence of excess charge on the floating gates, which is a satisfied condition without extra fabrication steps as a consequence of the gate and substrate connections to the top metal. A special programming technique for non-zero charge on the floating gates was not utilized in the discussed work, but a more sophisticated floating-gate device implementation as presented in [14] could be explored, which promises additional potential for compensation of inherent transistor threshold voltage offsets in the OTA's input differential pair.

The phase shifter in Fig. 3.2 creates an extra pole within the linearized architecture that the reference OTA does not have. This phase delay is roughly the same as the delay from the pole formed by R and C_o in the auxiliary path. In low-loss (high-Q) designs, the additional pole can affect the gain of integrators and the frequency response of biquad sections if $1/(RC_o)$ is not significantly larger than the operating frequency. A load compensation scheme based on [17] is explained in Appendix B for such situations.

Fig. 3.5 Error amplifier
circuit in the CMFB loop

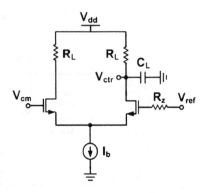

Identical standalone OTAs are included on the same die to obtain reference linearity measurements. The reference OTA also has a floating-gate input attenuation of 1/3 for fair performance comparison. In this way, the linearity benefit from the input attenuation is isolated from the architectural linearization in Fig. 3.2, and both OTAs have the same effective transconductance ($G_m/3$ in this case), but the linearization results in doubled power consumption. Since attenuation and feedback linearization techniques have known linearity and effective transconductance trade-offs, the circuit-level comparison is focused on the predistortion linearization scheme relative to a commensurate OTA with equal input attenuation factor. This baseline OTA in Fig. 3.4 was biased with $I_b = 0.95$ mA and $I_{b1} = 0.85$ mA, having an effective transconductance of 510 μA/V. The linearization does not require any design changes in this core OTA, but redesign of the OTA is an option if it is required to meet the same power budget after linearization, which is possible as long as OTA bandwidth reduction can be tolerated. Such a linearization under power constraint is disclosed in Appendix C.

Suppression of undesired common-mode signals and noise is vital for linearity at high frequencies. The common-mode feedback (CMFB) circuit should have high gain to accurately control the common-mode output voltage while maintaining a wide bandwidth to reject common-mode noise in the band of interest. The CMFB amplifier is shown in Fig. 3.5, where V_{ctr} is the control voltage applied to the OTA in Fig. 3.4. The addition of the compensation resistor R_z results in two zeros in the transfer function of the error amplifier, which helps to insure stability of the CMFB loop. The simulated AC response of the CMFB loop has a 51.9 dB low-frequency gain and a 424.9 MHz unity-gain frequency with 42.5° phase margin.

3.3.2 Proof-of-Concept Filter Realization and Application Considerations

A second-order G_m-C biquad filter was designed with attenuation-predistortion-linearized OTAs to verify that the methodology is suitable for filters with G_m-C integrator loops. Figure 3.6 shows the filter schematic and specifications.

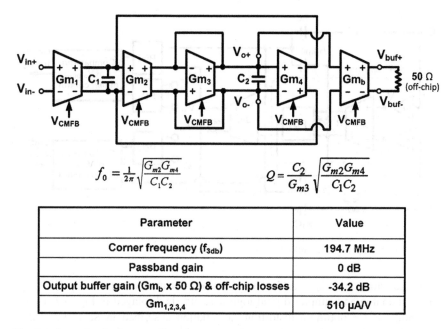

Fig. 3.6 Second-order lowpass filter diagram and design parameters

$$f_0 = \frac{1}{2\pi}\sqrt{\frac{G_{m2}G_{m4}}{C_1 C_2}} \qquad Q = \frac{C_2}{G_{m3}}\sqrt{\frac{G_{m2}G_{m4}}{C_1 C_2}}$$

Parameter	Value
Corner frequency (f_{3db})	194.7 MHz
Passband gain	0 dB
Output buffer gain (Gm_b x 50 Ω) & off-chip losses	-34.2 dB
$Gm_{1,2,3,4}$	510 µA/V

The lowpass output of the biquad was measured using another OTA as buffer to drive the 50 Ω input impedance of the spectrum analyzer.

The primary motivation for digital correction (Sect. 3.4) to enhance linearity performance with severe process variation is the compatibility with digitally-controlled receiver calibration approaches that involve the baseband filter. Practical implementation details for receivers with digital performance monitoring and calibration of analog blocks are reported in [18–21]. They incorporate accurate digital monitoring and I/Q mismatch correction in the digital signal processor (DSP) as well as a few analog observables that give some insights into the operating conditions, such as outputs from received signal strength indicators or DC control voltages of blocks. The possibility exists to generate and apply test tones at the input of an analog block and extract performance indicators from the output spectrum in the DSP, which contains distortion components. Conversely, calibration could also be performed by monitoring the bit error rate (BER) in the DSP from processing a special test sequence or customary pilot symbols at the beginning of receptions. Since linearity degradation impacts the BER, such a calibration could be computationally more efficient than calculating and analyzing the fast Fourier transform in the DSP. Regardless of the specific digital calibration algorithm, the digitally-controlled correction capability of the attenuation-predistortion linearization scheme can potentially enable filter linearity tuning in integrated receiver applications without the need for extra DACs.

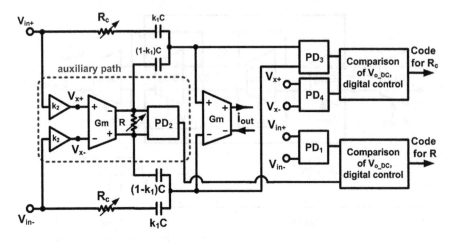

Fig. 3.7 Block diagram of an automatic linearity tuning scheme

An alternative automatic calibration that does not involve an on-chip DSP but dedicated analog and simpler digital logic circuitry is displayed in Fig. 3.7. From the conditions for optimum distortion cancellation described in Sect. 3.2.3, the gain of the auxiliary path must be equal to $k_2 G_m R$, which is unity in the discussed design example. This can be ensured by measuring the signal level at the input and output of the auxiliary OTA with peak or power detectors (PD_1, PD_2), and controlling the digital code of resistor R until the gain is unity. The simplest control algorithm would be to cycle through the codes that determine the value of R until the difference in the DC output voltages of PD_1 and PD_2 is minimized, which can be performed digitally by detecting the toggling instance at the output of a single comparator. At higher frequencies, the parasitic pole in the auxiliary path starts to affect the distortion cancellation, causing the signal level at the output of the auxiliary OTA to decrease with increasing frequency. Hence, the differential input signal to the main OTA at PD_3 increases as a result, which is shown in Fig. 3.8. By measuring this signal that is ideally equal to $V_x = k_2 \cdot V_{in}$ with PD_3, the value of the phase shift resistor R_c can be adjusted until the outputs of PD_3 and PD_4 are equal. This comparison can be completed with the same logic as for PD_1/PD_2, but is has to be done with an input signal at the maximum frequency at which high linearity is desired. The automatic tuning has not been implemented on the circuit-level, but simulations with different values of R_c showed that amplitude detection within 4.6% is required to detect R_c changes within 5% at 350 MHz, which is sufficient for IM3 higher than 70 dBc (Sect. 3.4). In differential gain measurements, PVT errors in the detectors are cancelled except for the errors from unavoidable mismatches between the two detectors. Errors from mismatches are less than 5% at 2.4 GHz in [22], and more accurate amplitude detection is achievable at lower frequencies. In [23] for example, differential on-chip amplitude measurements were conducted up to 2.4 GHz using detectors with 0.031 mm^2 die area and negligible loading of the signal path ($C_{in} < 15$fF).

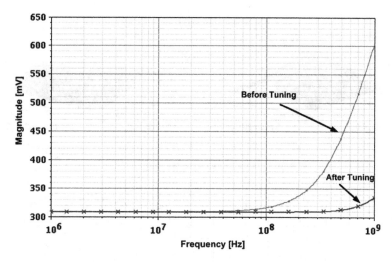

Fig. 3.8 Simulated AC amplitude at the input of the main OTA (PD$_3$ in Fig. 3.7)

3.4 Compensation for PVT Variations and High-Frequency Effects

Since the frequency compensation is based on equalization of phase shifts from RC time constants in the main and auxiliary paths, the optimum linearity point is subjected to PVT variations. Resistors R and R_c in Fig. 3.2 can be adjusted digitally to ensure high linearity. When implementing the attenuation ratios with matched capacitors, the variation of the resistors and transconductance mismatch between the auxiliary and main paths become the major sources of IM3 degradation. Figure 3.9 illustrates the technique's sensitivity to 20% variation of R_c and G_m based on the expression for IM3 in (3.6). In theory, the |IM3| (in dBc) without parameter variation is infinite. After introducing a numerical resolution constraint, the peak |IM3| is limited to around 95 dBc. Figure 3.9a reveals that G_m-mismatch results in more degradation than R_c variation at low frequencies, but at high frequencies variation of R_c becomes equally significant as evident from Fig. 3.9b. In general, less than ±10% mismatch of $G_m \times R$ and ±5% variation of R_c are required for theoretical |IM3| higher than 70 dBc. Under consideration of the trend towards increasing intra-die variability in modern CMOS processes, programmability of R and R_c is necessary to guarantee $G_m \times R$ gain and R_c values within these limits. The determination of the appropriate incremental resistor step size is elaborated next.

To obtain a practical assessment of the distortion cancellation sensitivity, the compensation resistor value and transconductance mismatch in the two paths were varied in circuit simulations using Spectre. The resulting |IM3| is plotted vs. deviation from the nominal design parameters in Fig. 3.10, showing an |IM3| better

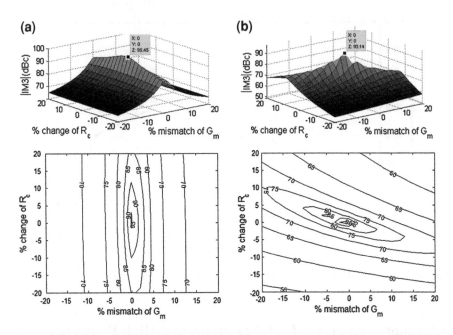

Fig. 3.9 Sensitivity of |IM3| (in dBc) to component mismatches. Calculated with Eq. 3.6: **a** 10 MHz signal frequency, **b** 200 MHz signal frequency

than 71 dBc for ±7.5% R_c-variation and |IM3| better than 71 dBc for ±3.3% R-variation in the presence of 10% G_m-mismatch. The reference OTA has |IM3| of 51 dBc. It is imperative for effective distortion cancellation to implement the resistor ladders with 3% steps, enabling digital correction of relatively small intra-die mismatches. To account for large absolute variations of parameters, the adequate resistor tuning range should be selected based on simulations under anticipated worst-case conditions. In the described work, simulations with process corner models and temperatures ranging from −40 to 100°C were conducted. Based on these simulation results, a conservative range from approximately 30 Ω to 2.2 kΩ (approximately 3–200% of the nominal value) and 6-bit resolution were chosen for the programmable resistors R_c and R (Fig. 3.2) in the prototype design.

3.5 Prototype Measurement Results

3.5.1 Standalone OTA

Table 3.1 summarizes the characterization results for the OTA by itself. Two 0.1 V_{p-p} (−16 dBm) tones with 100 kHz frequency separation and a combined voltage swing of 0.2 V_{p-p} were applied during IM3 measurements. The results in

Fig. 3.10 Simulated
sensitivity to critical
component variations and
mismatches. **a** |IM3| vs.
change in R_c at 350 MHz,
b |IM3| vs. R with 10%
transconductance mismatch
between main OTA and
auxiliary OTA at 350 MHz

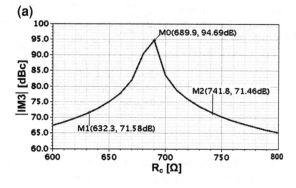

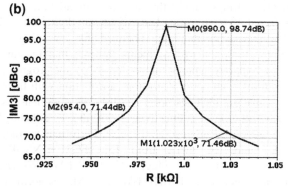

Fig. 3.11 demonstrate IM3 enhancement from -58.5 to -74.2 dB at 350 MHz
coupled with a rise in input-referred noise from 13.3 to 21.8 nV/$\sqrt{\text{Hz}}$ and twice
the power dissipation, while other performance parameters are not affected
significantly. The linearization decreased the SNR in 1 MHz BW from 74.5 to
70.2 dB, but allowed to improve the IM3 by 15.7 dB. Depending on the frequency
and settings of the switches, IM3 enhancement up to 22 dB was achieved with the
compensation resistor ladders having 6-bit resolution. If more linearity improvement is required, the resolution of the resistor ladders (R and R_c) in Fig. 3.2 can be
increased by adding more control bits or using a MOS in triode region as one of
the elements to obtain a series resistance that is closer to the optimum value for
distortion cancellation.

The IM3 from the two-tone tests of the reference and linearized OTAs around
350 MHz is plotted versus input peak-to-peak voltage in Fig. 3.12. This comparison demonstrates that the IM3 enhancement from the linearization scheme
requires weakly nonlinear operation. Even though the linearization effectiveness
decreases with increasing input signal swing, the IM3 improvement is still 11 dB
with 0.8 $V_{p\text{-}p}$ differential signal swing for the example design with 1.2 V supply.
Since the distortion cancellation exhibits the highest sensitivity to phase shifts at
high frequencies, the control code of the phase shift resistor R_c in Fig. 3.2 has been
changed from its optimum value. The resulting effect on the IM3 of the linearized

Table 3.1 Measured main parameters of the reference folded-cascode OTA

Parameter	Measurement
Transconductance (G_m)	510 µA/V
IM3 @ 50 MHz (V_{in} = 0.2 Vp–p)	−55.3 dB
Noise (input-referred)	13.3 nV/$\sqrt{Hz}$
Power with CMFB	2.6 mW
PSRR @ 50 MHz	48.9 dB
Supply	1.2 V

OTA at 350 MHz is plotted in Fig. 3.13, which validates that variable phase compensation is in fact required for optimum linearity performance. Two resistor ladder settings satisfy that the IM3 attenuation is more than 74 dB, hence the selected 3% step for the least significant digital bit in this design was appropriate. Together with the plot obtained by sweeping resistor R_c in simulations (Fig. 3.10a), the measurements indicate that the amount of IM3 improvement predominantly depends on the step size of the programmable resistor ladder, which promises even better distortion cancellation with finer resolution.

Table 3.2 includes noise and IM3 measurement results at various frequencies, demonstrating the effectiveness of the broadband linearization scheme with the associated input-referred noise. Performance trade-offs can be assessed with the figure of merit from [6]: FOM = NSNR + 10log(f/1 MHz), where NSNR = $SNR_{(dB)}$ + 10log[($IM3_N$/IM3)(BW/BW_N)(P_N/P_{dis})] from [7], the SNR is integrated over 1 MHz BW, the IM3 is normalized with $IM3_N$ = 1%, the bandwidth is normalized with BW_N = 1 Hz, and the power consumption is normalized with P_N = 1 mW. Experimental results are compared with previously reported architectures in Table 3.3. The OTA linearized with input attenuation-predistortion shows a competitive performance with respect to the state of the art. High linearity at high frequencies is realized in this design example, showing the potential of the technique.

3.5.2 Second-Order Lowpass Filter

Figure 3.14 shows the filter frequency response for the proof-of-concept biquad design in Fig. 3.6, and its linearity performance is plotted against frequency in Fig. 3.15. The IM3 of the filter is up to 8 dB worse than that of the standalone OTA. However, the measured filter IM3 includes approximately 2–3 dB degradation due to the nonlinearity of the output buffer, which was not de-embedded from the measurement results. By adjusting the resistor ladders with digital controls that are common for all OTAs, the filter achieves IM3 ≈ −70 dB up to 150 MHz for a 0.2 $V_{p–p}$ two-tone input. At 200 MHz, which is above the 194.7 MHz filter cutoff frequency, the IM3 is −66.1 dB, demonstrating the effectiveness of the broadband linearization due to compensation with the phase shifter.

Fig. 3.11 Measured linearity with 0.2 V_{p-p} input swing from two tones. (Each tone: 0.1 V_{p-p} (-16 dBm) on-chip after accounting for off-chip losses at the input). Displayed outputs: **a** reference OTA, **b** compensated OTA

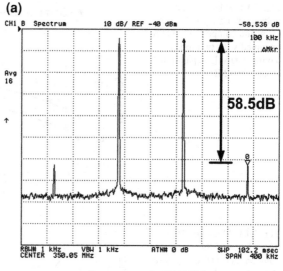

(a)

Uncompensated OTA IM3
(input: 0.2V$_{p-p}$@350MHz)

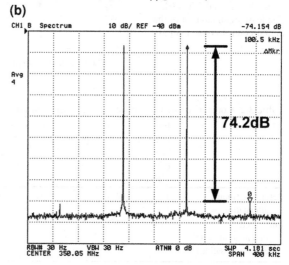

(b)

Compensated OTA IM3
(input: 0.2V$_{p-p}$@350MHz)

Figure 3.16 visualizes the measured IM3 with increasing input voltage up to 1.13 V peak–peak differential swing, which follows the expected trend. At 150 MHz, an IM3 of approximately -31 dB occurs with an input signal of 0.75 V_{p-p}.

Figure 3.17 illustrates the in-band third-order intermodulation intercept point (IIP3 = 14.0 dBm) and second-order intermodulation intercept point (IIP2 = 33.7 dBm) curves measured with two tones separated by 100 kHz around

Fig. 3.12 IM3 vs. input voltage swing for reference OTA and compensated OTA. Obtained with two tones having 100 kHz separation around 350 MHz

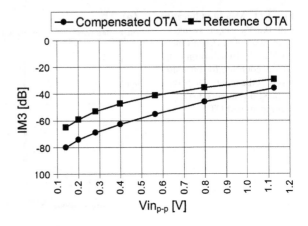

Fig. 3.13 Measured IM3 dependence of the compensated OTA on phase shift. Obtained with two test tones having 100 kHz separation around 350 MHz. The least significant bit of the digital control code changes R_c by ~3%

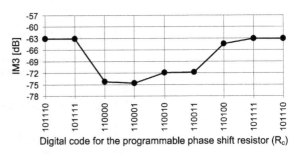

Table 3.2 Comparison of OTA linearity and noise measurements

| OTA type | Input-referred noise (nV/$\sqrt{Hz}$) | Power consumption (mW) | IM3 (V_{in} = 0.2 V_{p-p}) | | | Normalized |FOMI[a] (at 350 MHz) |
			50 MHz (dB)	150 MHz (dB)	350 MHz (dB)	
Reference (input attenuation = 1/3)	13.3	2.6	−55.3	−60.0	−58.5	56.7
Linearized (attenuation = 1/3 & compensation)	21.8	5.2	−77.3	−77.7	−74.2	64.3

[a] See Table 3.3 for details

150 MHz and 2 MHz, respectively. In broadband receiver applications with limited filtering in the RF front-end, the presence of numerous out-of-band interference signals results in intermodulation components within the desired signal band. Thus, high out-of-band linearity is desirable in addition to the baseband filter attenuation in order to minimize in-band distortion. This is one of the main motivations to employ OTAs with high linearity at high frequencies even for baseband filters with low cutoff frequencies. The out-of-band IIP3 plot in

Table 3.3 OTA comparison with prior works

	Reference [1][a]	Reference [8][a]	Reference [10]	Reference [9]	Reference [5][a]	This example
IM3	–	−47 dB	−70 dB	−60 dB	–	−74.2 dB
IIP3	−12.5 dBV	–	–	–	7 dBV	7.6 dBV
f_o	275 MHz	10 MHz	20 MHz	40 MHz	184 MHz	350 MHz
Input voltage	–	0.2 V_{p-p}	1.0 V_{p-p}	0.9 V_{p-p}	–	0.2 V_{p-p}
Power/transconductor	4.5 mW	1.0 mW	4 mW	9.5 mW	1.26 mW	5.2 mW
Input-referred noise	7.8 nV/$\sqrt{}$Hz	7.5 nV/$\sqrt{}$Hz	70.0 nV/$\sqrt{}$Hz	23.0 nV/$\sqrt{}$Hz	53.7 nV/$\sqrt{}$Hz	21.8 nV/$\sqrt{}$Hz
Supply voltage	1.2 V	1.8 V	3.3 V	1.5 V	1.8 V	1.2 V
Technology	65 nm CMOS	0.18 μm CMOS	0.5 μm CMOS	0.18 μm CMOS	0.18 μm CMOS	0.13 μm CMOS
FOM$_{(dB)}$[b]	87.5	92.9	96.1	99.1	100	105.6
Normalized |FOM|[c]	1.0	3.4	7.1	14.3	17.8	64.3

[a] Power/transconductor calculated from filter power. Individual OTA characterization results not reported in full

[b] FOM$_{(dB)}$ = 10log(f/1 MHz) + NSNR from [6]

NSNR = SNR$_{(dB)}$ + 10log[(IM3$_N$/IM3)(BW/BW$_N$)(P$_N$/P$_{dis}$)] from [7]

(SNR integrated over 1 MHz BW, normalization: IM3$_N$ = 1%, BW$_N$ = 1 Hz, P$_N$ = 1mW)

(IM3 in FOM for [1, 5] was calculated with: IM3$_{(dB)}$ = 2 × [Pin$_{(dBm)}$ − IIP3$_{(dB)}$])

[c] Normalized FOM magnitude relative to [1]

Normalized |FOM| = 10^(FOM$_{(dB)}$/10)/(10^(FOM$_{(dB)}$/10) of [1])

Fig. 3.14 Measured filter frequency response and linearity. **a** Transfer function with ∼34 dB total losses (input loss and output buffer attenuation). **b** IM3 with 0.2 V_{p-p} input swing from two tones, each 0.1 V_{p-p} (−16 dBm) on-chip after accounting for off-chip input losses

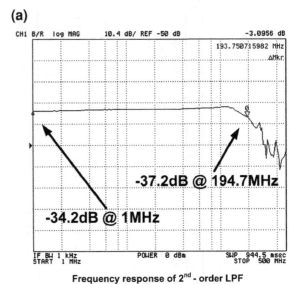

(a)

CH1 B/R log MAG   10.4 dB/ REF -50 dB   -3.0956 dB

193.750715982 MHz
ΔMkr

-37.2dB @ 194.7MHz

-34.2dB @ 1MHz

IF BW 1 kHz   POWER 0 dBm   SWP 944.5 msec
START 1 MHz   STOP 500 MHz

Frequency response of 2nd - order LPF

(b)

CH1 B Spectrum   10 dB/ REF -40 dBm   -69.682 dB

100 kHz
ΔMkr

Avg
5

69.7dB

RBW# 30 Hz   VBW 30 Hz   ATN# 0 dB   SWP 4.181 sec
CENTER 150 MHz   SPAN 400 kHz

IM3 of the LPF with compensated OTAs
(input: 0.2V_{p-p}@150MHz)

Fig. 3.18a confirms that the linearization scheme's effectiveness is preserved beyond the cutoff frequency. The slight degradation of the out-of-band IIP3 to 12.4 dBm is most likely due to the different phase shifts experienced by the 275 and 375 MHz test tones from the input to node 2 in the auxiliary path (Fig. 3.2). The digital control code for the phase shift resistor R_c of the OTAs in the filter was set to optimize linearity in the 195 MHz bandwidth, hence the linearity degradation due to the frequency difference of the out-of-band tones. The out-of-band IIP2 (Fig. 3.18b) is 30.4 dBm, which is 3.3 dB lower than the in-band

Fig. 3.15 Filter IM3 vs.
frequency measured with two
tones spaced by 100 kHz

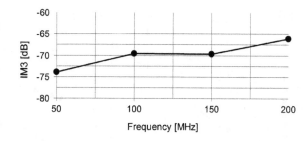

Fig. 3.16 IM3 vs. input
peak–peak voltage for the
linearized filter. Measured
with two test tones separated
by 100 kHz around 150 MHz

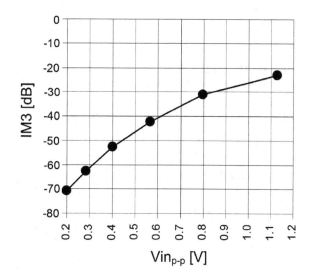

IIP2 due to suboptimum phase shifts at 375 MHz. Despite of that, the use of OTAs with high out-of-band linearity helps to reduce in-band distortion from out-of-band interferers in broadband scenarios.

Table 3.4 summarizes the filter's key performance parameters in contrast to other wideband lowpass filters. The 54.5 dB dynamic range integrated over the 195 MHz noise bandwidth is competitive with other works having similar power consumption per pole, most of which were implemented under less voltage headroom constraints than with the 1.2 V supply in the example design. The discussed linearization is independent of OTA topology, but the proof-of-concept design is comprised of a restrictive fully-differential OTA core in order to demonstrate the concept with a conventional topology. The last two columns in Table 3.4 indicate that the attenuation-predistortion linearization allows almost similar filter linearity performance (in-band IIP3 = 14.0 dBm with 1.2 V supply) by means of fully-differential OTAs as with the pseudo-differential OTAs in [28], in which an in-band IIP3 of 16.9 dBm was achieved with 1.8 V supply. Apart from linearity considerations, the optimizations involving power consumption, input-referred noise, power supply noise rejection, and CMRR depend on the

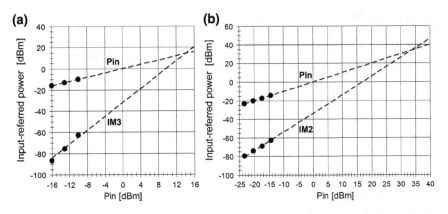

Fig. 3.17 Measured in-band intercept point curves for the filter. **a** IIP3 (two tones, $\Delta f = 100$ kHz around 150 MHz), **b** IIP2 (two tones, $\Delta f = 100$ kHz around 2 MHz)

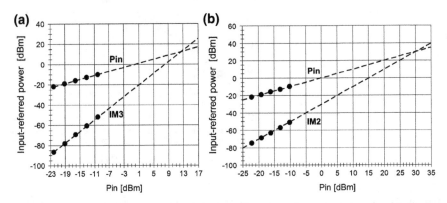

Fig. 3.18 Measured out-of-band intercept point curves for the filter. **a** IIP3 ($f_1 = 275$ MHz, $f_2 = 375$ MHz, $f_{IM3} = 100$ MHz), **b** IIP2 ($f_1 = 375$ MHz, $f_2 = 375.1$ MHz, $f_{IM2} = 100$ kHz)

application-specific constraints. According to the FOM comparison with the reference OTA in Table 3.2, the attenuation-predistortion linearization method improves linearity with justifiable power and noise trade-offs. Furthermore, the best dynamic range improvement with the technique can be achieved in bandpass designs, in which the noise is integrated over a narrow passband and the linearity improvement significantly reduces the power of the in-band distortion. The filter area on the die (Fig. 3.19) is approximately 0.5 mm^2 including the output buffer.

3.6 Summarizing Remarks

An attenuation-predistortion technique was described to linearize transconductance amplifiers in G_m-C filter applications over a wide frequency range and across PVT variations. The high-frequency linearity enhancement is based on Volterra series

Table 3.4 Comparison of wideband G_m-C lowpass filters

	Reference [1]	Reference [5]	Reference [24]	Reference [25]	Reference [26]	Reference [27]	Reference [28]	This example
Filter order	5	5	8	4	7	5	3	2
f_c (max.)	275 MHz	184 MHz	120 MHz	200 MHz	200 MHz	500 MHz	300 MHz	200 MHz
Signal swing	–	0.30 V_{p-p}	0.20 V_{p-p}	0.88 V_{p-p}	0.80 V_{p-p}	0.50 V_{p-p}	–	0.75 V_{p-p}
Linearity with max. $V_{in_{p-p}}$	–	HD3, HD5: <–45 dB	THD: –50 dB @ 120 MHz	THD: –40 dB @ 20 MHz	THD: –42 dB @ 200 MHz	THD: <– 40 dB @ 70 MHz	–	IM3: –31 dB[d] @ 150 MHz
In-band IIP3	–12.5 dBV (0.5 dBm)	7 dBV (20dBm)	–	–	–	–	3.9 dBV (16.9 dBm)	1.0 dBV (14.0 dBm)
In-band IIP2	–	–	–	–	–	–	19 dBV (32 dBm)	20.7 dBV (33.7 dBm)
Out-of-band IIP3	–8 dBV (5 dBm)	–	–	–	–	–	–	–0.6 dBV (12.4 dBm)
Out-of-band IIP2	15 dBV (28 dBm)	–	–	–	–	–	–	17.4 dBV (30.4 dBm)
Power	36 mW	12.6 mW	120 mW	48 mW	210 mW	100 mW	72 mW	20.8 mW
Power per pole	7.2 mW	2.5 mW	15 mW	12 mW	30 mW	20 mW	24 mW	10.4 mW
Input-Referred Noise	7.8 nV/$\sqrt{}$Hz	53.7 nV/$\sqrt{}$Hz[b]	–	–	–	–	5 nV/$\sqrt{}$Hz	35.4 nV/$\sqrt{}$Hz
Dynamic range	44 dB[a]	43.3 dB[c]	45 dB	58 dB	–	52 dB	–	54.5 dB[c]
Supply voltage	1.2 V	1.8 V	2.5 V	2 V	3 V	3.3 V	1.8 V	1.2 V
Technology	65 nm CMOS	0.18 μm CMOS	0.25 μm CMOS	0.35 μm CMOS	0.25 μm CMOS	0.35 μm CMOS	0.18 μm CMOS	0.13 μm CMOS

[a] Reported spurious-free dynamic range
[b] Calculated from 9.3 μV_{RMS} in 30 kHz BW
[c] Calculated from max. V_{P-P}, f_c, and input-referred noise density
[d] IM3 of –31 dB measured close to f_c ensures THD <–40 dB

Fig. 3.19 Die micrograph of the OTAs and filter in 0.13 μm CMOS technology. Reference OTA area: 0.033 mm², linearized OTA area: 0.090 mm²

analysis. Experimental results confirm the efficacy of the OTA linearization at high frequencies to obtain IM3 as low as -74 dB with 0.2 V_{in_p-p} at 350 MHz. Measurements of a biquad demonstrated that the linearization methodology is suitable for G_m-C filter applications requiring an overall IM3 ≤ -70 dB up to the cutoff frequency. The presented linearization approach is independent of the OTA architecture and robust due to the use of matched OTAs to cancel output distortion, resulting in an IM3 improvement of up to 22 dB. Compensation for PVT variations and high-frequency effects is based on digital adjustment of resistors without changing the bias conditions, which would affect other design parameters. Hence, the main OTA can be optimized for its target application.

References

1. V. Saari, M. Kaltiokallio, S. Lindfors, J. Ryynänen, K.A.I. Halonen, A 240 MHz low-pass filter with variable gain in 65-nm CMOS for a UWB radio receiver. IEEE Trans. Circuits Syst. Regul. Pap. **56**(7), 1488–1499 (2009)
2. M. Gambhir, V. Dhanasekaran, J. Silva-Martinez, E. Sánchez-Sinencio, A low power 1.3 GHz dual-path current mode Gm-C filter, in *Proceedings of IEEE Custom Integrated Circuits Conference (CICC)*, Sept 2008, pp. 703–706
3. R. Schoofs, M.S.J. Steyaert, W.M.C. Sansen, A design-optimized continuous-time delta-sigma ADC for WLAN applications. IEEE Trans. Circuits Syst. Regul. Pap. **54**(1), 209–217 (2007)
4. D. Healy, *Analog-to-Information (A-to-I) Receiver Development Program*, BAA 08-03 Announcement, Defense Advanced Research Projects Agency (DARPA), Microsystems Technology Office (MTO), Nov 2007

5. J.C. Rudell, O.E. Erdogan, D.G. Yee, R. Brockenbrough, C.S.G. Conroy, B. Kim, A 5th-order continuous-time harmonic-rejection GmC filter with in situ calibration for use in transmitter applications, in *IEEE International Solid-State Circuits Conference (ISSCC) Digest of Technical Papers*, Feb 2005, pp 322–323

6. A. Lewinski, J. Silva-Martinez, A high-frequency transconductor using a robust nonlinearity cancellation. IEEE Trans. Circuits Syst. Express Briefs 53(9), 896–900 (2006)

7. E.A.M. Klumperink, B. Nauta, Systematic comparison of HF CMOS transconductors. IEEE Trans. Circuits Syst. Express Briefs 50(10), 728–741 (2003)

8. S. D'Amico, M. Conta, A. Baschirotto, A 4.1 mW 10 MHz fourth-order source-follower-based continuous-time filter with 79 dB DR. IEEE J. Solid-State Circuits 41(12), 2713–2719 (2006)

9. T.Y. Lo, C.-C. Hung, A 40 MHz double differential-pair CMOS OTA with −60 dB IM3. IEEE Trans. Circuits Syst. Regul. Pap. 55(1), 258–265 (2008)

10. J. Chen, E. Sánchez-Sinencio, J. Silva-Martinez, Frequency-dependent harmonic-distortion analysis of a linearized cross-coupled CMOS OTA and its application to OTA-C filters. IEEE Trans. Circuits Syst. Regul. Pap. 53(3), 499–510 (2006)

11. W. Huang, E. Sánchez-Sinencio, Robust highly linear high-frequency CMOS OTA with IM3 below −70 dB at 26 MHz. IEEE Trans. Circuits Syst. Regul. Pap. 53(7), 1433–1447 (2006)

12. D. Yongwang, R. Harjani, A +18 dBm IIP3 LNA in 0.35 µm CMOS, in *IEEE International Solid-State Circuits Conference (ISSCC) Digest of Technical Papers*, Feb 2001, pp. 162–163

13. M. Mobarak, M. Onabajo, J. Silva-Martinez, E. Sánchez-Sinencio, Attenuation- predistortion linearization of CMOS OTAs with digital correction of process variations in OTA-C filter applications. IEEE J. Solid-State Circuits 45(2), 351–367 (2010)

14. R. Chawla, F. Adil, G. Serrano, P.E. Hasler, Programmable Gm-C filters using floating-gate operational transconductance amplifiers. IEEE Trans. Circuits Syst. Regul. Pap. 54(3), 481–491 (2007)

15. S. Maas, *Nonlinear Microwave and RF Circuits* (Artech House, Boston, 2003)

16. E. Rodriguez-Villegas, H. Barnes, Solution to trapped charge in FGMOS transistors. Electron. Lett. 39(19), 1416–1417 (2003)

17. A.P. Nedungadi, R.L. Geiger, High-frequency voltage-controlled continuous time lowpass filter using linearized CMOS integrators. Electron. Lett. 22, 729–731 (1986)

18. D. Kaczman, M. Shah, M. Alam, M. Rachedine, D. Cashen, L. Han, A. Raghavan, A single-chip 10-band WCDMA/HSDPA 4-band GSM/EDGE SAW-less CMOS receiver with DigRF 3G interface and +90 dBm IIP2. IEEE J. Solid-State Circuits 44(3), 718–739 (2009)

19. H. Darabi, J. Chiu, S. Khorram, H.J. Kim, Z. Zhou, H.-M. Chien, B. Ibrahim, E. Geronaga, L.H. Tran, A. Rofougaran, A dual-mode 802.11b/Bluetooth radio in 0.35 µm CMOS. IEEE J. Solid-State Circuits 40(3), 698–706 (2005)

20. I. Vassiliou, K. Vavelidis, T. Georgantas, S. Plevridis, N. Haralabidis, G. Kamoulakos, C. Kapnistis, S. Kavadias, Y. Kokolakis, P. Merakos, J.C. Rudell, A. Yamanaka, S. Bouras, I. Bouras, A single-chip digitally calibrated 5.15–5.825 GHz 0.18 µm CMOS transceiver for 802.11a wireless LAN. IEEE J. Solid-State Circuits 38(12), 2221–2231 (2003)

21. Y.-H. Hsieh, W.-Y. Hu, S.-M. Lin, C.-L. Chen, W.-K. Li, S.-J. Chen, D.J. Chen, An auto-I/Q calibrated CMOS transceiver for 802.11g. IEEE J. Solid-State Circuits 40(11), 2187–2192 (2005)

22. A. Valdes-Garcia, R. Venkatasubramanian, R. Srinivasan, J. Silva-Martinez, E. Sánchez-Sinencio, A CMOS RF RMS detector for built-in testing of wireless transceivers, in *Proceedings of IEEE VLSI Test Symposium*, May 2005, pp. 249–254

23. A. Valdes-Garcia, R. Venkatasubramanian, J. Silva-Martinez, E. Sánchez-Sinencio, A broadband CMOS amplitude detector for on-chip RF measurements. IEEE Trans. Instrum. Meas. 57(7), 1470–1477 (2008)

24. G. Bollati, S. Marchese, M. Demicheli, R. Castello, An eighth-order CMOS low-pass filter with 30–120 MHz tuning range and programmable boost. IEEE J. Solid-State Circuits 36(7), 1056–1066 (2001)

25. A. Otin, S. Celma, C. Aldea, A 40–200 MHz programmable 4th-order Gm-C filter with auto-tuning system, in *Proceedings of 33rd European Solid-State Circuits Conference (ESSCIRC)*, Sept 2007, pp. 214–217
26. S. Dosho, T. Morie, H. Fujiyama, A 200 MHz seventh-order equiripple continuous-time filter by design of nonlinearity suppression in 0.25 μm CMOS process. IEEE J. Solid-State Circuits **37**(5), 559–565 (2002)
27. S. Pavan, T. Laxminidhi, A 70–500 MHz programmable CMOS filter compensated for MOS nonquasistatic effects, in *Proceedings of 32nd European Solid-State Circuits Conference (ESSCIRC)*, Sept 2006, pp. 328–331
28. K. Kwon, H.-T. Kim, K. Lee, A 50–300 MHz highly linear and low-noise CMOS Gm-C filter adopting multiple gated transistors for digital TV tuner ICs. IEEE Trans. Microwave Theory Tech. **57**(2), 306–313 (2009)

Chapter 4
Multi-Bit Quantizer Design for Continuous-Time Sigma-Delta Modulators with Reduced Device Matching Requirements

Abstract Future wireless devices will require extensive connectivity to accommodate several services, which means that the receivers must cover broader frequency bands. Therefore, on-chip analog-to-digital converters (ADCs) in multi-standard receivers not only demand increased signal-to-quantization-noise-ratio, but also more bandwidth for the conversion of the analog signals into the digital domain. This chapter briefly introduces a lowpass continuous-time $\Sigma\Delta$ ADC architecture that was developed for next generation broadband receiver applications. Rather than using multiple signal levels, a multi-bit digital-to-analog converter (DAC) realization based on a feedback signal with time-varying pulse duration was employed. This approach alleviates nonlinearity problems associated with typical multi-bit DACs. The chapter also contains an in-depth description of the corresponding 3-bit quantizer architecture with multi-phase clocking. The reference levels for this quantizer are adjustable to compensate for process variations after fabrication if the application necessitates fine resolution.

4.1 Background

The 3-bit quantizer examined in this chapter was specifically designed as part of a continuous-time $\Sigma\Delta$ modulator [1], which is an analog-to-digital converter (ADC) that uses oversampling and filtering to achieve quantization noise-shaping

---

This chapter includes portions reprinted with permission, from "A 25MHz bandwidth 5th-order continuous-time lowpass sigma-delta modulator with 67.7dB SNDR using time-domain quantization and feedback", C.-Y. Lu, M. Onabajo, V. Gadde, Y.-C. Lo, H.-P. Chen, V. Periasamy, and J. Silva-Martinez, *IEEE J. Solid-State Circuits*, vol. 45, no. 9, pp. 1795–1808, Sept. 2010., © 2010 IEEE.

M. Onabajo and J. Silva-Martinez, *Analog Circuit Design for Process Variation-Resilient Systems-on-a-Chip*, DOI: 10.1007/978-1-4614-2296-9_4, © Springer Science+Business Media New York 2012

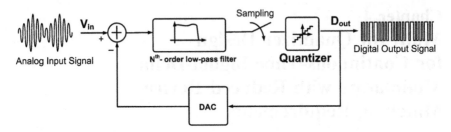

Fig. 4.1 Simplified diagram of a continuous-time $\Sigma\Delta$ modulator

to obtain an effective number of bits (signal-to-quantization-noise ratio) significantly higher than the quantizer in the loop (e.g. a 12-bit ADC). Such an ADC is visualized in Fig. 4.1 just to show the quantizer's location in the loop, where the most conventional quantizer architecture is a flash ADC. Details regarding the operation and design of typical continuous-time $\Sigma\Delta$ modulators are outside of the scope of this book, but they can be found in [2].

4.1.1 State of the Art Continuous-Time $\Sigma\Delta$ ADCs

Various wireless standards such as WiMAX have been developed over the years due to the high demand for faster data rate in portable wireless communications, which has pushed baseband bandwidths up to a few tens of megahertz. When high-resolution continuous-time lowpass $\Sigma\Delta$ ADC architectures are selected for emerging products because of their efficiency, a wide bandwidth is essential in multi-standard applications to accommodate receiver bandwidth requirements. A popular way to improve the signal-to-noise-and-distortion ratio (SNDR) over wide bandwidth without increasing the sampling frequency is to use a multi-bit quantizer and a multi-bit feedback digital-to-analog converter (DAC) [3]. With this approach, the noise-shaping gain required in the loop filter can be relaxed due to the reduced quantization noise associated with the multi-bit quantizer. Even though multi-bit architectures have been successfully utilized in multi-MHz bandwidth designs, the "digital friendly" advantages of the 1-bit architecture are typically compromised with the multi-bit solution. In particular, the feedback DAC nonlinearity significantly affects the ADC performance because it directly adds error to the filter input signal and it is not noise-shaped. Dynamic element matching (DEM) and data weighted averaging (DWA) techniques have been proposed to tackle this problem [4–6]. However, the additional power and complexity of DEM methods is not permissible in some applications. In a previous work [7], the thresholds of the comparators in a 9-level quantizer were shuffled rather than performing DAC element rotations, which shortens the delay for the mismatch-shaping realization. The feasibility of this method has been demonstrated in a modulator having 82 dB SNDR over 10 MHz bandwidth with a 5th-order loop filter. In general, the shaping of the mismatch error provided by DEM/DWA techniques is less effective for designs with low oversampling ratio (OSR) and high conversion speeds due to excessive loop delay. On the contrary, the line of attack in the work discussed in this chapter is to prevent DAC element matching issues altogether by using a multi-bit single-element DAC. This strategy, on the other hand, necessitates fast and accurate digital timing circuitry, which is a trade-off whose attractiveness parallels the ongoing technology scaling.

Recent practical works have incorporated a digital-intensive time-based multi-bit quantizer [8] and quantizer/DAC combination [9] in the modulator architecture, achieving 72 and 60 dB SNDR over 10 and 20 MHz bandwidths, respectively. Since scaling of CMOS process technologies provides an advantageous environment for high-speed digital timing control but perilous conditions for analog device matching in the DAC/quantizer, the time-based approaches and pulse-width modulation (PWM) feedback DACs are promising solutions for future technologies. The recent simulation results for the designs in [10] provide further insights into the effectiveness of this design methodology. In anticipation of increasing process variations, the approach taken in the described example involves a 3-bit quantizer and a single-element DAC that realizes 3-bit feedback via time-based operation (generation of a PWM waveform). Hence, the need for DAC unit element matching or DEM/DWA techniques is eliminated. However, time-based approaches require strict control over the timing signals and clock jitter to attain high SNDR. The main trade-off is that the DAC linearity depends on the mismatches between the clock phases for the PWM waveform generation rather than unit element mismatches as in conventional multi-bit DACs.

4.1.2 Quantizer Design Trends

Figure 4.2 displays a typical 3-bit flash quantizer, in which the input signal V_{in} is compared to seven reference voltage levels (obtained with a resistor ladder) using seven comparators (C_1–C_7). For high-speed operation, the comparators are often comprised of preamplifiers followed by latches. The quantization occurs in one clock period and it usually yields thermometer code as output, which can be converted to the desired digital output code with an encoder. With regards to PVT variations, a relevant condition is that the resistors (R) must be matched in the layout to avoid shifts in the reference voltage levels. Moreover, the input-offset voltages of the comparators are subjected to PVT variations, in particular through the worsening threshold voltage variations (Table 2.1). Compensating for these variations and the resulting offsets that cause ADC nonlinearity errors is an ongoing research topic to which many solutions have been proposed over the past decades. Similar to transceiver system calibration approaches (Sect. 2.2.5), recently proposed methods involve calibration control in the digital domain in combination with programmable circuit element through the use of switches. In [11] for example,

Fig. 4.2 Conventional 3-bit
flash quantizer

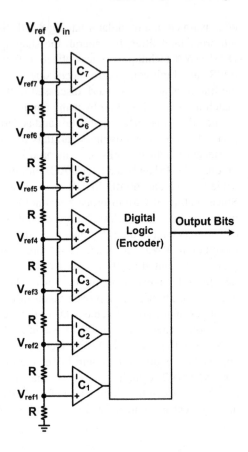

additional resistors are included in the reference voltage ladder to generate extra
voltage levels between the ideal references. The best combination of references is
selected with switches and a digital control scheme in order to compensate for offsets
from process variations, improving the effective number of bits of the flash ADC
from 3 to 5.6. Another recently proposed digital calibration technique employs
programmable load resistors in the differential preamplification stage within the
comparators in order to make adjustments that counteract random offsets [12].
To maintain compatibility with such digital calibration methods, the quantizer
architecture introduced in this chapter has been constructed to allow reference
voltage tuning without affecting components that are directly in the signal path.

Traditional two-step flash analog-to-digital converter (ADC) architectures are a
subset of subranging ADCs that typically consist of a sample-and-hold (S/H), a
most-significant bit(s) (MSB) ADC, a digital-to-analog converter (DAC), a gain
block, and a least-significant bit(s) (LSB) ADC [13]. As an example, the adapted
block diagram of the two-step ADC described in [13] is displayed in Fig. 4.3,
which utilizes two DACs. Conceptually, the operation is as follows: After the input
signal is sampled by the S/H circuit, the MSBs are resolved using a fixed reference

Fig. 4.3 The two-step ADC principle

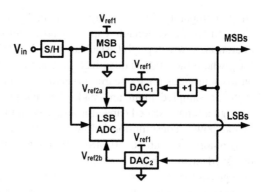

voltage range (V_{ref1}). Next, DAC_1 generates the upper reference voltage (V_{ref2a}) for the decision with the LSB ADC by incrementing the quantized MSB value by one in the digital domain. The lower range for the LSB decision is set with DAC_2, which directly converts the quantized MSB into an analog voltage (V_{ref2b}). With the selected reference voltage subrange, the LSB ADC performs a fine quantization of the sampled input voltage. Such a two-step flash approach has the advantage that the output bits from two low-resolution ADCs can be combined to obtain more precision, reducing the number of comparators that a conventional flash ADC would require for the same resolution. Hence, multi-step quantization can be used to lower area and power consumption when a delay of multiple clock cycles or clock phases can be tolerated.

In the past years, several alternative quantizer architectures have been proposed to optimize the operation by taking advantage of technology scaling for enhanced performance at higher conversion speeds, reducing power consumption, and improving compatibility with digital CMOS processes. However, design challenges also arise from adverse effects in deep-submicron technologies such as reduced gains from lower transistor output impedances, design with limited voltage headroom, reduced transistor linearity, and increased PVT variations as well as intra-die variability. As a result, recent works involved quantizer design trade-offs that exploit the advantage of modern CMOS processes while avoiding the drawbacks. For instance, the folding flash ADC in [14] is comprised of 16 instead of 31 (conventional flash) comparators for 5-bit resolution to decrease the power consumption. In addition, the folding topology in [14] circumvents the use of amplifiers in 90 nm CMOS technology, which increases its merit with regards to scaling and integration. With the availability of fast-switching devices, successive approximation ADCs are not constrained to low-speed operation anymore as demonstrated by realizations with low- to medium-resolution at medium- to high-speed [15–17]. The 6-bit 600 MS/s ADC in [15] exemplifies how asynchronous processing can be utilized to shorten the comparison cycles when employing multiple comparisons to resolve the bits from MSB to LSB sequentially. In [15], the asynchronous successive approximations are performed with a single comparator by weighing the input against a reference that is dynamically changed with a switchable capacitor array before each comparison. With similar operation, a two-step 7-bit ADC having a 150 MS/s conversion rate is described in [16], where the MSB is quantized first and the remaining bits are determined with an asynchronous binary-search procedure. In [17], the successive approximations with the sampled input are made via charge-sharing that occurs while cycling through a binary-scaled capacitor array. With the comparator being the only active block, power consumption below 0.7 mW was achieved with the 9-bit ADC at conversion rates up to 50 MS/s. When a multi-bit lowpass $\Sigma\Delta$ modulators is designed with a high oversampling ratio, then the sample-to-sample voltage changes of the slow-varying input signal are small. Therefore, only a small number of comparators connected to the reference voltages above and below the current signal level are required in consecutive conversions with a conventional flash architecture. This characteristic can be exploited to reduce the number of comparators by either shifting the references associated with the reduced number of comparators or by shifting the input signal prior to the comparison. For instance, the lowpass $\Sigma\Delta$ modulator with 104 MHz sampling frequency and 2 MHz bandwidth in [18] uses a tracking ADC with 3 comparators in lieu of a conventional 4-bit flash quantizer that would require 15 comparators, which shows how quantizer operation can be optimized for its application in a specific $\Sigma\Delta$ modulator architecture.

4.1.3 Quantizer Design Considerations for the $\Sigma\Delta$ Modulator Architecture

When designing high-resolution lowpass multi-bit $\Sigma\Delta$ ADCs in modern CMOS technologies with rising process variations, the linearity performance of the feedback DAC at the ADC input becomes a limiting factor for the overall performance because its nonlinearity errors are not noise-shaped by the loop dynamics. The quantizer discussed in this chapter has been designed as part of a project in which an alternative multi-bit feedback approach was explored by constructing a $\Sigma\Delta$ ADC architecture that does not rely on unit element matching in the front-end DAC. Instead, the $\Sigma\Delta$ ADC employs an inherently linear single-element PWM DAC that is controlled via multi-phase clock signals. The general aim of this approach is to circumvent analog device matching requirements by relying more on well-timed digital operations. Figure 4.4 depicts the fully-differential 5th-order lowpass $\Sigma\Delta$ modulator with a sampling frequency of 400 MHz for 25 MHz signal bandwidth. A 5th-order quasi-linear phase inverse Chebyshev lowpass filter with 49 dB pass-band gain is employed, which consists of two cascaded active-RC 2nd-order lowpass sections and a lossy integrator with sufficient linearity. The summing amplifier couples all feedforward paths of the filter to the quantizer input. A level-to-PWM converter translates the multi-bit signal into a

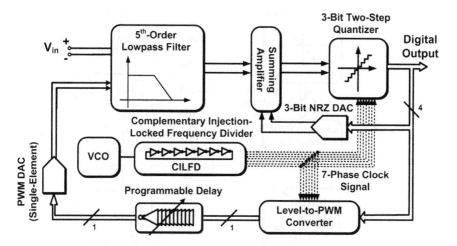

Fig. 4.4 Block diagram of the 5th-order continuous-time modulator

time-domain digital PWM signal such that only a 1-bit current-steering DAC is required for global feedback with 3-bit equivalence. This realization avoids performance degradation originating from current mismatch linked to conventional multi-bit DACs at the modulator input. A 2.8GHz inductor-capacitor tank voltage-controlled oscillator (VCO) and a ring oscillator type complementary injection-locked frequency divider (CILFD) [19] produce low-jitter clock signals at 400 MHz with seven evenly distributed phases (Φ_1–Φ_7) for the digital logic of the quantizer and the level-to-PWM converter. The nonidealities of the local 3-bit non-return-to-zero (NRZ) DAC feeding into the quantizer input are noise-shaped by the modulator loop, making this DAC design less critical. Hence, a standard 3-bit DAC was chosen for the local feedback.

Due to the requirements of wide bandwidth and high resolution, combinations of multi-bit quantizer and DACs generating multi-level signals are commonly employed. In conventional current-steering DACs, the amplitude levels of the feedback current at the input of the loop filter are generated by adding the outputs of the appropriate number of unit element current sources for the quantizer output code. Device mismatches from process variations generate out-band noise that folds into the frequency range of interest as well as in-band harmonic distortion components that degrade the modulator's SNDR. Solutions such as noise-shaping dynamic element matching (DEM) [4], tree-structure DEM [5], and the data weighted averaging technique [6] were proposed to reduce the DAC linearity degradation from mismatch. However, improvements in wideband ADCs are usually limited due to restrictions on loop delay and increased noise levels from the randomization procedure. In this chapter's design, a single-element DAC having an output waveform with variable pulse width per sampling period generates a 3-bit charge injection feedback as shown in Fig. 4.5. Since only one inherently linear single-bit DAC

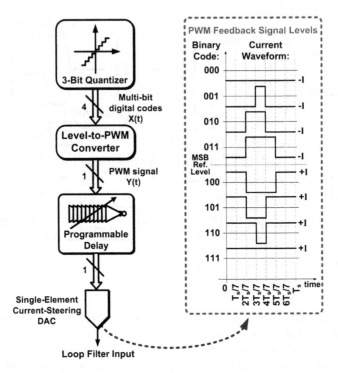

Fig. 4.5 Feedback path with 3-bit quantizer and PWM DAC

produces different feedback charge levels at the loop filter input, the current mismatch issue of multi-amplitude DACs is avoided. A level-to-PWM converter is implemented in the feedback path to convert the digital codes from the 3-bit quantizer to time-domain PWM signals compatible with the 1-bit DAC having time-varying output pulses of current amplitude $\pm I$. The PWM DAC output pulse shapes are arranged as symmetric as possible within a clock period to minimize the power of potential aliasing tones [10]. These pseudo-symmetric high and low amplitude levels of the single-element DAC during one clock period are also visualized in Fig. 4.5 together with their binary equivalent codes.

The main drawback of employing multi-phase time-domain signals is increased sensitivity to jitter noise because of larger and more frequent DAC output transitions compared to a conventional NRZ DAC. The peak signal-to-jitter-noise ratio (SJNR) of the modulator can be analytically estimated for a feedback pulse shape with [2]:

$$SJNR_{peak} = 10 \cdot \log_{10} \left(\frac{T_s^2 \cdot OSR}{2 \cdot \sigma_y^2 \cdot \sigma_\beta^2} \right) \tag{4.1}$$

where $OSR = 1/(2 \cdot \mathrm{BW} \cdot T_s)$, σ_β is the clock jitter standard deviation, and σ_y is the standard deviation of $[y(n) - y(n-1)]$; with $y(n)$ being the nth combined digital output of the modulator. The SJNR of the modulator with level-to-PWM converter was evaluated in comparison to a conventional 3-bit modulator [20], showing that the simulated SJNR limit of the PWM DAC with $\sigma_\beta \approx 0.5$ ps is 5 dB lower than that of a conventional 3-bit NRZ DAC at 400 MHz. Furthermore, the worst-case clock jitter requirement for SNDR > 68 dB with the discussed modulator is $\sigma_\beta < 0.54$ ps [20].

The nonlinearity of the PWM DAC due to static timing mismatches can be assessed from a feedback charge error comparison relative to the conventional 3-bit DAC. Figure 4.6 visualizes the worst-case peak-to-peak charge errors for each code, which are resultants of static mismatch ΔI_i for each current cell in the conventional DAC and static timing error ΔT_j of clock phase Φ_j in the PWM DAC. ΔT_j originates from static CILFD mismatches and unequal propagation delays due to routing parasitics, but it does not accumulate in the inverter chain because each stage is locked to the VCO signal. The ideal feedback charge per code is identical for both DACs. Notice that the errors depend on mismatches in up to seven unit elements of the conventional DAC, but only up to two timing phases with the PWM scheme in which two edges define the area under the pulse regardless of the deviations that the phases in between have. Assuming equal mismatches ($\Delta I_i = \Delta I$, $\Delta T_j = \Delta T$) yields worst-case errors of $\pm 7\Delta I \cdot T_s$ and $\pm 2\Delta T \cdot I$ for conventional and PWM DACs, respectively. Letting $\delta_{\%I} = \Delta I/(I/7)$ and $\delta_{\%T} = \Delta T/(T_s/7)$ be the percent standard deviations of the mismatches in each case, the worst-case accumulated errors are $\Delta Q_{conv.-worst} = \pm 7\delta_{\%I} \cdot (I/7) \cdot T_s$ and $\Delta Q_{PWM-worst} = \pm 2\delta_{\%T} \cdot I \cdot (T_s/7)$. Monte Carlo simulations including delay mismatches in all clock phases showed that $\delta_{\%T} = 0.16\%$ as a result of the synchronizing effect from the injection-locking. Since $\delta_{\%I}$ is typically 0.5% with good layout practices for a standard DAC, the anticipated worst-case linearity error of the PWM DAC is favorably lower. Assuming that two timing mismatches are accumulated in the case of the PWM-based ADC, all mismatches in the conventional realization are accumulated, and errors are un-correlated in both cases; the induced 3rd-order harmonic distortion (HD3) ratio can be estimated as derived in [20]:

$$\frac{HD3_{PWM}}{HD3_{conventional}} \simeq \left(\sqrt{\frac{2}{N}} \right) \left(\frac{\delta_{\%T}}{\delta_{\%I}} \right) \qquad (4.2)$$

where N is the number of DAC levels. For $N = 7$ and the aforementioned distributions, the linearity of the described PWM DAC theoretically outperforms the conventional DAC by 15.3 dB according to (4.2). It is important to note that this estimated improvement is based on the timing mismatch prediction from Monte Carlo simulations of this particular clock generation circuitry, and that nonidealities such as supply noise and ground bounce should be minimized to avoid PWM DAC linearity degradations due to timing errors in the digital circuitry.

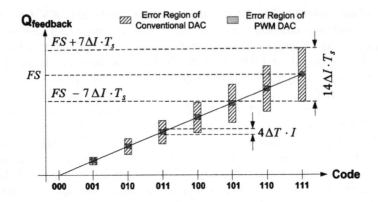

$|\Delta Q|$	000	001	010	011	100	101	110	111
Conventional DAC	0	$\Delta I_1 \cdot T_s$	$\sum_{i=1}^{2} \Delta I_i \cdot T_s$	$\sum_{i=1}^{3} \Delta I_i \cdot T_s$	$\sum_{i=1}^{4} \Delta I_i \cdot T_s$	$\sum_{i=1}^{5} \Delta I_i \cdot T_s$	$\sum_{i=1}^{6} \Delta I_i \cdot T_s$	$\sum_{i=1}^{7} \Delta I_i \cdot T_s$
PWM DAC	0	$(\Delta T_4 + \Delta T_5) \cdot I$	$(\Delta T_3 + \Delta T_5) \cdot I$	$(\Delta T_3 + \Delta T_6) \cdot I$	$(\Delta T_3 + \Delta T_6) \cdot I$	$(\Delta T_3 + \Delta T_5) \cdot I$	$(\Delta T_4 + \Delta T_5) \cdot I$	$2\Delta T_1 \cdot I$

Fig. 4.6 Relative 3-bit DAC linearity error comparison: conventional vs. PWM

4.2 3-Bit Two-Step Current-Mode Quantizer Architecture

4.2.1 Quantizer Design

As illustrated in Fig. 4.7, the quantizer utilizes the seven on-chip clock phases to control four sequential comparison instances (τ_1–τ_4), which cuts the number of comparators from seven to four with respect to a typical 3-bit flash ADC. The two-step process makes the MSB available after the first step, creating timing margin for the digital control logic that sets up the PWM DAC. Successive approximations during the second step resolve the remaining bits that are processed by the level-to-PWM converter. As a result, and similar to the combination of the PWM generator and TDC in [9], the 1-bit DAC is driven by a PWM waveform. However, in the approach presented here, successive approximations are employed for comparison with the input signal rather than generation of a continuous ramp. Since this successive algorithm only has one MSB and three LSB quantization steps, the comparison to discrete reference levels is a simple alternative that also gives the option to calibrate each level individually if necessary.

4.2.1.1 Decision Timing

The quantizer operates as follows with regards to the topology in Fig. 4.7 and corresponding timing diagram in Fig. 4.8. The differential input signal V_{in} is sampled with a S/H circuit by the 400 MHz master clock having a period T_s, and

Fig. 4.7 Single-ended equivalent block diagram of the quantizer

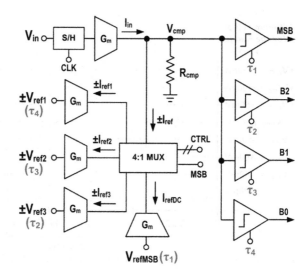

then it is converted to current I_{in} via a transconductance stage (G_m). First, the MSB is resolved after τ_1 seconds by comparing I_{in} to the current from V_{refMSB} applied to an identical G_m stage. Depending on the timing control bits (CTRL) and the MSB decision, a multiplexing configuration (MUX) is utilized to compare I_{in} to current I_{ref} derived from the appropriate differential reference voltage ($\pm V_{ref1}...\pm V_{ref3}$) during each subsequent instant (τ_2–τ_4). The order of the subranging comparisons and output bits was chosen based on the timing needs in the multi-phase DAC control circuitry because larger signal magnitudes require DAC feedback pulse changes early in the next clock cycle. Comparison resistor (R_{cmp}) converts the difference in currents into a positive or negative voltage. A binary result of the current-mode comparison is stored using a latched comparator for each of the four decisions. The tabular inset in Fig. 4.8 lists the output codes corresponding to the input ranges.

4.2.1.2   Circuit-Level Design Considerations

Figure 4.9 displays the schematic of the quantizer core in which the current-mode comparisons are made. All devices with the same names are equal-sized and matched in the layout. The simplified S/H circuit represents a transistor-level implementation with gate-bootstrapping [21], and the AND gates effectively function as time-controlled MUX. After the S/H operation, the differential input voltage is converted to current by the transistor pair (M_n) and mirrored 1:1 by pair M_p. The other M_n transistors convert the differential reference voltages to currents for successive comparisons, where the difference current flows through the load resistors R_{cmp} to generate $V_{cmp} = V_{cmp+} - V_{cmp-}$. In this fully-differential circuit, $V_{refMSB} = 0$ V (MSB decision) level is obtained by applying the DC voltage

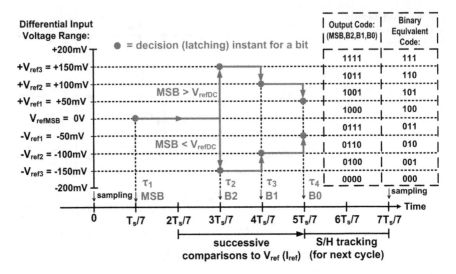

Fig. 4.8 Timing of the successive quantization decisions and output code words. The *arrows* show the two possible sequences based on the MSB value

V_{refCM} that is equivalent to the 1.1 V common-mode level at the input of the quantizer to both transistors in one of the branches for comparison with the input signal. The other differential reference voltages listed below Fig. 4.9 were selected to span the 400 mV$_{p-p}$ full-scale swing at the quantizer input. For each reference current step, the polarity of this differential voltage is resolved by the latched comparator.

Polysilicon resistors (R_{BW}) in Fig. 4.9 extend the bandwidth of the current mirrors [22] for high-frequency operation according to:

$$BW_{mirror} = \frac{1}{2\pi}\sqrt{g_{mp}/(R_{BW} \cdot C_{gsp}^2)} \tag{4.3}$$

where g_{mp} and C_{gsp} are the transconductance and gate-source capacitance of M$_p$, correspondingly. With $R_{BW} = 330\ \Omega$, the simulated 3 dB bandwidth of the current mirrors is 3.36 GHz, which is sufficiently high to prevent it from becoming the factor that limits the comparison speed. More critical is that speed performance is ensured by selecting the value of resistors R_{cmp} such that the RC time constant formed with parasitic capacitance C_p at the comparison nodes (V_{cmp+}, V_{cmp-}) does not impose limitations. After switch M$_{sw}$ closes to compare the current from the input signal with the corresponding reference in each comparison cycle, the difference current $I_{cmp} = I_{cmp+} - I_{cmp-}$ will cause a step response at the input of the latches ($V_{cmp} = V_{cmp+} - V_{cmp-}$). With a first-order model, this step response can be expressed as

$$V_{cmp}(S) = \frac{2 \cdot R_{cmp}}{1 + s \cdot R_{cmp} \cdot C_p} \times \frac{I_{cmp}}{s} \tag{4.4}$$

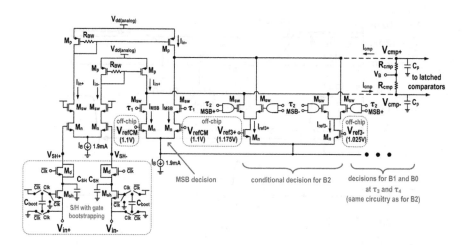

Fig. 4.9 Simplified schematic of the current-mode quantizer core circuitry. Reference voltages $\pm V_{ref3} = \pm 150$ mV $= \pm (V_{ref3+} - V_{ref3-}) = \pm (1.175$ V $- 1.025$ V$)$ and $\pm V_{refMSB} = 0$V $= V_{refCM} - V_{refCM} = 1.1 - 1.1$ V are shown. The other references are: $\pm V_{ref2} = \pm 100$ mV $= \pm (1.15$ V $- 1.05$ V$)$, $\pm V_{ref1} = \pm 50$ mV $= \pm (1.125$ V $- 1.075$ V$)$

where $s = j\omega$ and C_p is the cumulative parasitic capacitance at the comparison node from transistors M_p, M_{sw}, input devices of the four latches, as well as routing parasitics. Taking the inverse Laplace transform of (4.4) gives the transient response during each comparison phase:

$$V_{cmp(t)} = 2 \cdot I_{cmp} \cdot \left(R_{cmp} - R_{cmp} \cdot e^{-t/(R_{cmp} \cdot C_p)} \right) \qquad (4.5)$$

Figure 4.10 displays one sampling clock cycle of the simulated transient behavior at the comparison node, where the polarity (delineated by marker A on the $V_{cmp+} - V_{cmp-} = 0$ V line) of the differential voltage is latched on the falling edge of the shown timing signals that correspond to τ_1–τ_4 in Fig. 4.8. The latching instants are labeled with arrows, resulting in an output code of (MSB, B2, B1, B0) $= (1, 0, 0, 0)$ for this example quantization cycle. Note, $V_{cmp(t)}$ settles within 5% of its final value after approximately three $R_{cmp}C_p$ time constants. In this design, R_{cmp} is 405 Ω and C_p is approximately 250 fF, resulting in a theoretical time constant of 100 ps. Nevertheless, it is only critical that V_{cmp} is larger than the resolution of the latch that resolves whether V_{cmp} is positive or negative. This zero-crossing event must occur sufficiently early to allow pre-charging of the nodes inside the activated latch by its preamplifier within the $T_s/7$ comparison time window of the LSBs. If the aforementioned zero-crossing is delayed due to the large parasitic capacitance (C_p) or insufficient preamplification prior to latching, then false decisions could occur. Hence, the timing and signal amplitude at this comparison node is the most significant factor affecting the quantizer resolution. Note, other factors such as the switch turn-on delay (of M_{sw} in Fig. 4.9), finite rise/fall times of the control signals, delay variations of the control signals, clock jitter, and kickback from the latches also

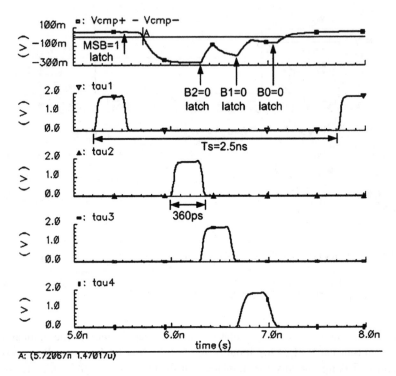

Fig. 4.10 Simulated example of the quantization timing. From *top to bottom*: transient voltage at the comparison node ($V_{cmp} = V_{cmp+} - V_{cmp-}$), signals τ_1–τ_4 that trigger latching on the falling edge

impact the decision accuracy and cause the deviation of the V_{cmp} signal waveform in Fig. 4.10 from the ideal sequence of step responses.

The clocked comparators connected to V_{cmp} in Fig. 4.7 are implemented with the fully-differential circuit shown in Fig. 4.11. In the tracking phase, Φ_{LA} is low and bias current I_B is steered into the preamplifier stage consisting of input transistor M_1 and load resistor R_{L1}. To save power, the bias current is reused in the latch phase (high Φ_{LA}) when it flows into M_{LA1}. Devices M_2, R_{L2}, M_{LA2} form a second preamplification and latch stage, but this stage is controlled by the phase-reversed latch signal to hold the decision for almost one clock period (T_s). Transistors M_7–M_{10} form a self-biased differential amplifier [23] which creates a rail-to-rail output during the long latch phase to drive the subsequent CMOS inverter (M_P, M_N).

The preamplifier and first latch stage also play an important role in the quantizer operation, impacting the overall resolution and speed that can be achieved. First of all, the input transistor M_1 in Fig. 4.11 should be as small as permissible to avoid introduction of excessive capacitance at the output node of the current-mode comparator core. The associated trade-off with small dimensions is increased input offset, which should be assessed via statistical simulations. Secondly, the band-width of the preamplifier must high to avoid delay. In this design, its first pole is

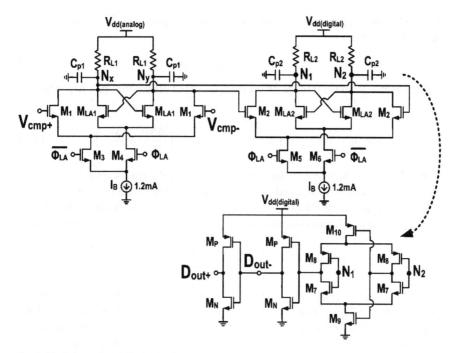

Fig. 4.11 Schematic of the latched comparator

around 3.5 GHz with $R_{LI} = 570\ \Omega$ and $C_{pI} = 80$ fF including routing parasitics. With sufficient preamplifier bandwidth margin, the most critical timing constraint is the propagation delay t_{LAI} of the first latch, which can be estimated with the expression below obtained by substituting the preamplifier gain $(g_{mI}R_{LI})$ into the equation from [24].

$$t_{LA1} = \frac{C_{p1}}{g_{mLA1}} \cdot \ln\left(\frac{V_{OH} - V_{OL}}{2 \cdot g_{m1}R_{L1}(V_{cmp+} - V_{cmp-})}\right) \qquad (4.6)$$

In (4.6), g_{mI} and g_{mLAI} are the transconductances of M_1 and M_{LA1} in Fig. 4.11; and $V_{OH} - V_{OL}$ is the output voltage difference at nodes N_x and N_y between high and low logic levels after latching, which is 1.4 V in this design.

4.2.2 Process Variations

4.2.2.1 Mismatch Analysis

Since transistor dimensions should be small for optimum speed, the input offset of the first latch stage (Fig. 4.11) must be assessed carefully during the design. Neglecting the charge injection errors, this input offset can be expressed for the latch under investigation by utilizing the general expression for a latched comparator from [25]:

$$V_{off} = V_{off1} + V_{off2}/(g_{m1}R_{LA1}) \tag{4.7}$$

The offset V_{off1} in (4.7) is the offset from the input differential pair M_1, which is [26]:

$$V_{off1} = \Delta V_{T1} + (1/2) \cdot (V_{gs1} - V_{T1}) \cdot \left(\frac{\Delta R_{L1}}{R_{L1}} + \frac{\Delta \beta_1}{\beta_1} \right) \tag{4.8}$$

where ΔV_{T1} is the threshold voltage mismatch, V_{gs1} is the gate-source voltage, $\Delta \beta_1$ is the W/L mismatch of M_1, and ΔR_{L1} is the preamplifier load resistor mismatch.

From [27], the latch offset V_{off2} in (4.7) also depends on its threshold voltage (ΔV_{TLA1}) variation, device dimensions, and gate-source voltage overdrive (V_{gsLA1}):

$$V_{off2} = \Delta V_{TLA1} + \frac{V_{gsLA1} - V_{TLA1}}{2} \left(\frac{\Delta W_{LA1}}{W_{LA1}} - \frac{\Delta L_{LA1}}{L_{LA1}} \right) + \frac{\Delta Q}{C_p} \tag{4.9}$$

where W_{LA1} and L_{LA1} are the width and length of M_{LA1}, and ΔQ is the charge injection error. Charge injection from control signals should be minimized by using small-sized switching devices because it can cause decision errors. A comparator reset or compensation technique might be required if the application mandates better resolution. In this analysis, charge injection error is omitted for simplicity and to maintain a focus on the expressions that show how transistor sizes and bias conditions can be optimized for enhanced resolution with timing constraints and device mismatches, which both have more severe impact on the performance of the presented quantizer topology. Based on the analysis in [28], the following equations can be used as guidelines during the design of the first latch stage in Fig. 4.11 in order to minimize the variances (σ^2) corresponding to the above offset voltages:

$$\sigma^2_{off} = \sigma^2_{off1} + \left(\frac{1}{g_{m1}R_{LA1}} \right)^2 \cdot \sigma^2_{off2} \tag{4.10}$$

$$\sigma^2_{off1} = \frac{A^2_{VT1}}{W_1 L_1} + \frac{(V_{gs1} - V_{T1})^2}{4} \cdot \left(\frac{A^2_{RL1}}{W_{RL1} L_{RL1}} + \frac{A^2_{\beta M1}}{W_1 L_1} \right) \tag{4.11}$$

$$\sigma^2_{off2} = \frac{A^2_{VTLA1}}{W_{LA1} L_{LA1}} + \frac{(V_{gsLA1} - V_{TLA1})^2}{4} \cdot \frac{A^2_{\beta LA1}}{W_{LA1} L_{LA1}} \tag{4.12}$$

where A_x represents the process-dependent mismatch constant for parameter x with units of: (units of x) $\times$ μm. The above expressions reveal the trade-off between input offset voltage and speed because offset reduction requires large devices with minimal V_{gs}, which increases the parasitic capacitances and reduces the effective transconductances of the transistors at high frequencies.

Fig. 4.12 Latched comparator Monte Carlo simulation without device matching. Histograms (100 runs) for critical offsets in the first comparator stage (Fig. 4.11): **a** ΔV_{T1} (threshold voltage difference of transistor pair M_1), **b** input offset voltage (at gates of transistor pair M_1)

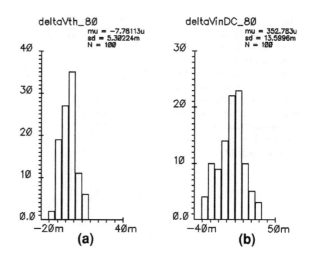

Monte Carlo simulations were performed to verify that the static offset voltages of the latched comparator and current-mode core are expected to cause errors less than 10% of the 50 mV quantization step, which are noise-shaped by the $\Sigma\Delta$ modulator. Figure 4.12 displays the histograms from 100 Monte Carlo runs at 80°C assuming that none of the devices are matched in layout. The threshold voltage mismatch (ΔV_{T1}) of transistor pair M_1 and the overall input offset (at $V_{cmp+/-}$ in Fig. 4.11) have standard deviations of 5.3 and 13.6 mV, respectively. In this simulation result from the complete quantizer circuit, the overall input offset at $V_{cmp+/-}$ is affected by the mismatches of the circuitry that impact the DC voltage at $V_{cmp+/-}$ in Fig. 4.9, including the comparison resistors (R_{cmp}). To determine the impact of this offset on the quantizer resolution, $V_{cmp} = V_{cmp+} - V_{cmp-}$ (Figs. 4.9, 4.11) has to be related to the measurable difference of $V_{in} - V_{refx}$, where $V_{in} = V_{in+} - V_{in-}$ (assuming negligible sampling errors) and $V_{refx} = V_{refx+} - V_{refx-}$ is an arbitrary differential reference voltage in Fig. 4.9. The input current and the subtracted current at the comparison node depend on V_{in}, V_{ref} and the transconductance g_{mn} of M_n. Since, the difference current flows into R_{cmp} to generate V_{cmp}, it can be shown that the following expression relates V_{cmp} to $V_{in} - V_{refx}$:

$$V_{in} - V_{refx} = \frac{V_{cmp}}{g_{mn} R_{cmp}} \tag{4.13}$$

which is $V_{in} - V_{refx} = V_{cmp}/2.63$ in the example design since $g_{mn} = 6.5$ mA/V and $R_{cmp} = 405\ \Omega$. Using Eq. 4.13 to refer the 13.6 mV input offset of the latched comparator to the quantizer input results in 5.2 mV. Such an input offset contribution from the latched comparator alone would be too high for the intended application, which is why the devices in Fig. 4.11 with identical labels were matched in the layout. Hence, the Monte Carlo simulations were repeated with correlation coefficients of 0.95 for the matched transistors and of 0.97 for the

Fig. 4.13 Latched comparator Monte Carlo simulation with device matching. Histograms (100 runs) for critical offsets in the first comparator stage (Fig. 4.11): **a** ΔV_{T1} (threshold voltage difference of transistor pair M_1), **b** input offset voltage (at gates of transistor pair M_1)

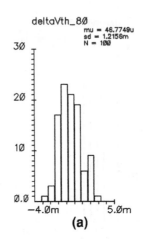

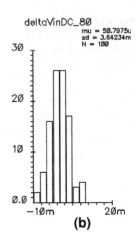

(a)

(b)

matched polysilicon resistors. The results in Fig. 4.13 show that the standard deviations ΔV_{T1} of transistor pair M_1 and the input offset voltage of the latched comparator reduce to 1.2 and 3.6 mV, respectively. After referring the latched comparator's input offset to the quantizer input based on (4.13) as before, the estimated input offset standard deviation becomes 3.6 mV/2.63 = 1.37 mV with device matching. *Thus, about 95% of the chips are expected to have an input-referred offset voltage below 2.7 mV (within two standard deviations assuming a Gaussian distribution) due to latched comparator mismatches.*

An input offset variation evaluation was also conducted for the differential pairs M_n in the current-mode comparator core (Fig. 4.9). All transistors with identical names in the core are also matched with a common-centroid layout, and Fig. 4.14 shows the histogram of the threshold voltage difference obtained from 100 Monte Carlo runs using the same correlations as defined in the latched comparator simulations. The estimated standard deviation is 0.97 mV for each differential pair M_n, which is the approximate input offset under the assumption that the errors from the matched current mirrors (M_p in Fig. 4.9) are not significant. Since the output currents of two differential pairs are compared in this circuit, the effective input offset voltage is found by combining the variances:

$$V_{off(core)} = \sqrt{\sigma_{Mn(input)}^2 + \sigma_{Mn(reference)}^2} = \sqrt{2 \cdot V_{off\_Mn}^2} \qquad (4.14)$$

where $\Sigma_{Mn(input)}$ is the standard deviation of the input offset voltage of the differential pair M_n in the input signal path and $\sigma_{Mn(reference)}$ is the standard deviation of the input offset voltage of the equal-sized reference differential pair by which the comparison current is generated. From Fig. 4.14 and Eq. 4.14, the estimated combined input-referred offset voltage during a comparison in the current-mode core is 1.4 mV. *Hence, about 95% of the chips are expected to have an input-referred current-mode comparator core offset below 2.8 mV, which is additional static error because this offset is directly at the input.*

Fig. 4.14 Quantizer core
Monte Carlo simulation with
device matching. Histogram
(100 runs) for the threshold
voltage difference of pairs M_n
in Fig. 4.9

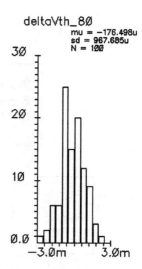

In summary, the latched comparator and current-mode comparator core input
offsets are expected to create a *combined static input-referred inaccuracy of less
than 5.5 mV with likelihood of 95%*. However, this error can be compensated by
tuning the reference voltages in Fig. 4.9 as demonstrated by a simulation in the
next subsection.

4.2.3 Simulation Results and Technology Scaling

4.2.3.1 Post-layout Simulations

Figure 4.15 shows the layout of the quantizer, which was designed in 0.18 μm
CMOS technology and embedded in the ΣΔ modulator. The quantizer's 0.39 mm^2
die area includes the bias and timing generation circuitry that generates the control
signal from the seven clock phases provided by the on-chip complementary
injection-locked frequency divider.

The simulation testbench included models for pad parasitics, bonding wire
inductances, and 100 fF capacitance as rough estimate for the effect of the
package. The output bit transitions during a ramp test with an input between −200
and 200 mV is shown in Fig. 4.16. From it, the transition levels with typical
device models were verified to be within approximately ±5 mV of the ideal
values, which is 10% of the 50 mV quantization step size. One level deviates by
7.6 mV from the ideal 150 mV. From system-level simulations of the continuous-
time ΣΔ ADC in Matlab, it was determined that up to ±10 mV reference level
shifts are permissible to achieve a signal-to-quantization noise ratio better than
72 dB.

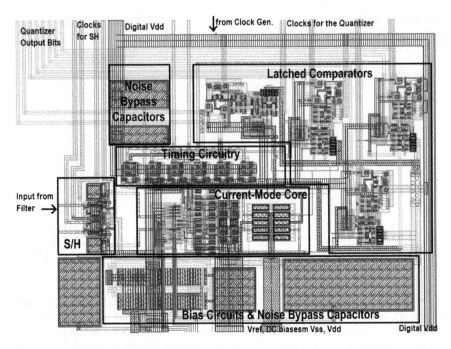

Fig. 4.15 Quantizer layout (0.18 μm CMOS technology). Area of quantizer and timing circuitry: 750 × 520 μm

Figure 4.17 displays the differential nonlinearity (DNL) and the integral non-linearity (INL) corresponding to the transition levels from the post-layout simulation. The adjustable reference voltages in Fig. 4.9 offer a way to alleviate the effects of PVT variations. As an example, Fig. 4.18 visualizes this feature for the −150 mV transition level of the 0.18 μm design, which can be shifted ±30 mV by adjusting V_{ref3}.

4.2.3.2 Technology Scaling

The operation of the current-mode quantizer architecture relies heavily on switching transistors and digital auxiliary circuitry. Hence, performance improvements can be expected in technologies with devices that have a high unity gain frequency (high-f_T). To verify this hypothesis, the quantizer and control circuitry were re-designed with UMC 90 nm CMOS technology and 1 V supply voltage, and then simulated with the identical setup as the 0.18 μm design. The dimensions of the components in the quantizer core (Fig. 4.9) are given in Table 4.1 for both designs, which shows that the active area with UMC 90 nm technology was reduced by more than four times. But, over half of the quantizer layout area (Fig. 4.15) in 0.18 μm technology consists of metal routing area and

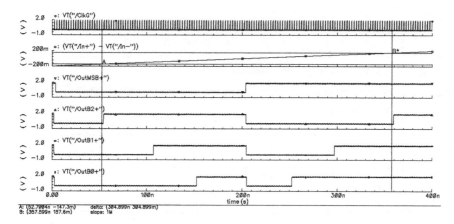

Fig. 4.16 Output bit transitions with an input ramp from −200 to 200 mV. *Top-to-bottom*: clock signal, input ramp, bits from Fig. 4.8: MSB, B2…B0. Quantization transition levels: −147.3, −94.8, −49.9, 2.7, 50.2, 95.1, 157.6 mV

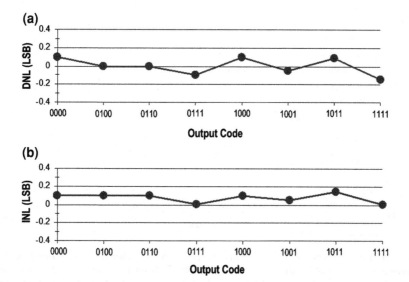

Fig. 4.17 Quantizer post-layout simulations: **a** DNL, **b** INL

capacitors to filter out noise at critical nodes. Since the requirements for routing and passives do not change significantly, an area reduction of up to 25% in the 90 nm process is a more reasonable estimate.

Table 4.2 provides a comparison of the most important quantizer properties, showing that the resolution in 90 nm is only reduced by 2 mV. This reduction is partially due to the limited voltage headroom with a 1 V supply, which causes

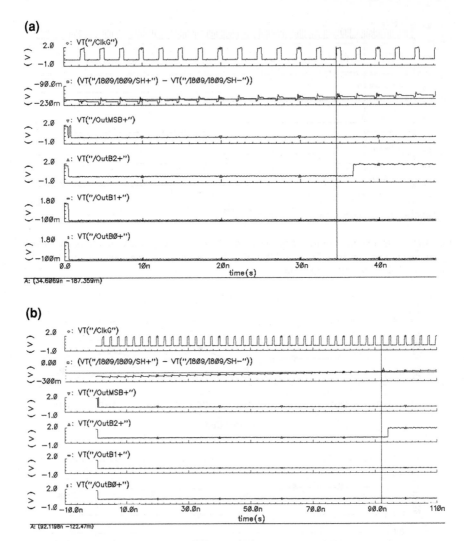

Fig. 4.18 Tuning range of the −150 mV transition level (schematic simulations). **a** Bit transition at −187.4 mV with $V_{ref3+} - V_{ref3-} = 180$ mV, **b** bit transition at −122.5 mV with $V_{ref3+} - V_{ref3-} = 120$ mV. *Top-to-bottom*: clock signal, input (after S/H), bits from Fig. 4.8: MSB, B2...B0

inaccuracies with large input signals because the devices operate at the edge of their intended regions of operations. Nevertheless, design optimizations through the use of non-minimum device dimensions could be explored to improve the resolution of the 90 nm design at the expense of an increase in power consumption.

A significant reduction in power was possible for the 90 nm design, in which the quantizer core consumes only 0.5 mW. On the contrary, this core consumed

Table 4.1 Component parameters in the quantizer core (Fig. 4.9)

Device	Jazz 0.18μm CMOS design (W/L dimensions or parameter)	UMC 90nm CMOS design (W/L dimensions or parameter)
M_n	28/0.2 μm	10/0.36 μm
M_p	56/0.18 μm	20/80 nm
M_{sw}	21/0.18 μm	16/80 nm
M_{B1}, M_{B2} for I_B (current mirror)	800/1 μm	110/1 μm
R_{cmp}	405 Ω	633 Ω
R_{BW}	333 Ω	1.4 kΩ
I_B	1.9 mA	0.25 mA
V_{refMSB}	1.1 V	0.5 V
V_{ref1+}/V_{ref1-}	1.125/1.075 V	0.525/0.475 V
V_{ref2+}/V_{ref2-}	1.150/1.050 V	0.550/0.450 V
V_{ref3+}/V_{ref3-}	1.175/1.025 V	0.575/0.425 V
Supply voltage	1.8 V	1 V

Table 4.2 Key quantizer performance parameters

	Jazz 0.18 μm CMOS	UMC 90 nm CMOS
Resolution	±5 mV	±7 mV
Static power: quantizer core	6.8 mW	0.5 mW
Static power: latched comparators	4 × 4.3 mW	4 × 0.3 mW
Layout area	750 × 520 μm (actual area for core, logic, routing)	estimate: ∼500 × 500 μm (∼1/4 of active area, similar passives/routing)
Clock frequency	400 MHz	400 MHz

6.8 mW in the initial 0.18 μm design. The power savings were enabled by the facts that the quantizer operation mainly depends on switching speeds and on the amount of parasitic capacitance from all devices connected to nodes V_{cmp+} and V_{cmp-} in Fig. 4.9. At these nodes, the parasitic capacitances form RC time constants with resistors (R_{cmp}) that limit the speed of the comparison. On the whole, less current is required to perform the comparisons with a 400MHz clock rate due to the smaller dimensions and higher ratio of transconductance to parasitic capacitance (i.e. higher f_T) in 90 nm technology.

4.2.4 ADC Chip Measurements with Embedded Quantizer

As mentioned earlier, the two-step current-mode quantizer has been designed for a $\Sigma\Delta$ modulator chip that was fabricated by our research group. Due to the complexity of the system, the test chip and printed circuit board were not equipped

Fig. 4.19 Die
microphotograph (2.6 mm^2
area excluding pads and ESD
circuitry)

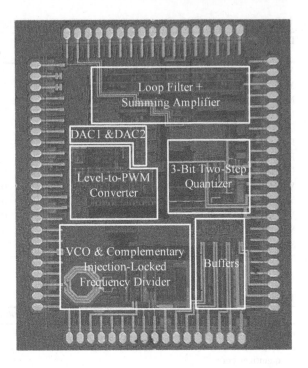

with sufficient inputs and outputs to characterize the individual blocks. A brief
overview of system-level measurements is presented in this subsection to demonstrate the 3-bit quantizer's functionality and that the block-level requirements
have been met to achieve the targeted system performance. Figure 4.19 displays
the die microphotograph of the multi-phase continuous-time 5th-order lowpass $\Sigma\Delta$
modulator fabricated in Jazz Semiconductor 0.18 μm 1P6M CMOS technology,
which was assembled in a QFN-80 package. It occupies a total area of 2.6 mm^2,
including the VCO and CILFD but excluding pads and electrostatic discharge
(ESD) protection circuitry. The four output bit streams of the 3-bit quantizer were
captured with a 4-channel oscilloscope synchronized at 400 Msamples/s prior to
post-processing in Matlab.

Figure 4.20 shows the output spectrum of the modulator with an input of
−2.2 dBFS at 5 MHz. Based on the noise bandwidth of 6.1 kHz during the
measurement, the average noise floor is around −145 dBFS/Hz and the peak SNR
is 68.5 dB in 25 MHz bandwidth. The third-order harmonic distortion (HD3) in
this case is 78 dB below the test tone, which demonstrates the high linearity
properties of both the loop filter and the PWM DAC/quantizer feedback scheme.
The peak SNDR including the harmonic tones in the 25 MHz bandwidth is
67.7 dB. The measured SNR and SNDR for different input signal powers are
plotted in Fig. 4.21, in which the 69 dB dynamic range (DR) is annotated.

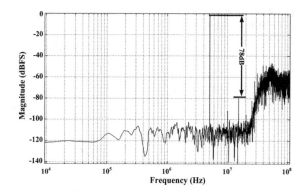

Fig. 4.20 Measured output spectrum of the ΣΔ modulator. A −2.2 dBFS input tone was applied at 5.08 MHz

Table 4.3 provides a summary of the modulator specifications. The linearity performance (IM3) was characterized by injecting two tones with 2 MHz separation, each having a power of −5 dBFS. Excluding the VCO, the power budget is 44 mW for the modulator core, 2.5 mW for the locked ring oscillator, and 1.5 mW due to clock buffers. Table 4.4 shows a comparison between the discussed modulator architecture and recently reported modulators based on the following figure-of-merit (*FoM*):

$$FoM = \frac{Power}{2^{ENOB} \cdot (2 \cdot BW)} \tag{4.15}$$

where *ENOB* is the effective number of bits and *BW* is the bandwidth. Although fabricated in an economical technology, the achieved 444 fJ/bit *FoM* of the described modulator core is competitive with the current state of the art. In addition, a *FoM* improvement is anticipated if the solution is exported to deep submicron technologies, which would lower the quantizer power (see Table 4.2) and level-to-PWM converter power as a result of more efficient switching circuitry.

4.3 Summarizing Remarks

A two-step current-mode quantizer was described in this section. The architecture was constructed for application within a ΣΔ modulator loop, and it incorporates characteristics that are aligned with present-day quantizer design trends. First, successive approximations controlled by multiple clock phases are used to reduce the number of required comparators in comparison to the classical flash quantizer architecture. Since switching operations become more efficient as technology scaling progresses, the successive comparison scheme in the quantizer core helps to take advantage of the speed benefits in modern CMOS technologies. Second, the quantizer has easily adjustable reference voltage levels, allowing it to be part of a system-level calibration technique as discussed in Sect. 2.2.5. In such a scenario,

Fig. 4.21 Measured SNR
and SNDR vs. input signal
power

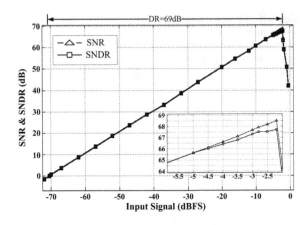

Table 4.3 Measured ΣΔ ADC performance

Technology	Jazz 0.18 μm CMOS
Power supply	1.8 V
Clock frequency	400 MHz
Bandwidth	25 MHz
Peak SNR/SNDR[a] @ 25MHz bandwidth	68.5/67.7 dB
SFDR	78 dB
IM3 (−5 dBFS per tone)	<−72 dB
Dynamic range	69 dB
Power consumption	48 mW
Area without pads and ESD protection	2.6 mm^2

[a] Includes total in-band distortion power and noise

the on-chip voltage references at the high-impedance input gates in the quantizer core (Fig. 4.9) can be generated with a low-power on-chip DAC.

With regards to the ΣΔ modulator application for which the quantizer was designed, the utilization of time-based processing methods within the continuous-time ΣΔ modulator shifts more operations into the digital realm, improving the system's robustness, scalability, and potential for power savings. A 5th-order continuous-time lowpass ΣΔ modulator using 3-bit time-domain quantization and feedback has been demonstrated in a 0.18 μm CMOS process. Nonlinearities from element mismatch of traditional multi-level DACs are circumvented because the 3-bit PWM feedback is realized with an inherently linear single-element DAC. Since low-jitter clocks are essential in time-based continuous-time ΣΔ modulators, the required jitter performance is accomplished by means of an injected-locked clock generation technique which provides 400 MHz clock signals with seven phases. The measured peak SNDR of the modulator with 25 MHz bandwidth is 67.7 dB, while the SFDR and DR are 78 and 69 dB, respectively. Its power consumption is 48 mW from a 1.8 V supply. Approximately 56% of this power is

Table 4.4 Comparison with previously reported lowpass ΣΔ ADCs

Reference	Technology (nm CMOS)	f_s (MHz)	BW (MHz)	Filter order	Peak SNDR (dB)	Power (mW)	FoM (fJ/bit)
[7] ISSCC 2008	180	640	10	5	82	100[†]	487
[8] JSSC 2008	130	950	10	2	72	40[a]	500
[9] ISSCC 2009	65	250	20	3	60	10.5[b]	319
[29] ISSCC 2007	90	340	20	4	69	56[c]	608
[30] ISSCC 2008	90	420	20	4	70	28[b]	271[d]
[31] JSSC 2006	130	640	20	3	74	20[b]	122
[32] ISSCC 2009	130	900	20	3 (+1 digital)	78.1	87[a]	330
This example	180 nm CMOS	400	25	5	67.7	48[a] (44[b])	484[a] (444[b])

[a] Includes clock generation circuitry; [b] For modulator circuitry only; [c] Includes digital calibration of RC spread and noise cancellation filter. [d] Discrete-time modulator (would require anti-aliasing filter for comparable blocker rejection)

dissipated in the quantizer and the level-to-PMW converter, which mainly contain circuits based on high-frequency switching. Technology scaling is expected to significantly enhance the efficiency of the presented modulator architecture via power reduction in the digital circuitry, especially in the quantizer.

References

1. C.-Y. Lu, M. Onabajo, V. Gadde, Y.-C. Lo, H.-P. Chen, V. Periasamy, J. Silva-Martinez, A 25 MHz bandwidth 5th-order continuous-time lowpass sigma-delta modulator with 67.7dB SNDR using time-domain quantization and feedback. IEEE J. Solid-State Circuits **45**(9), 1795–1808 (2010)
2. J.A. Cherry, W.M. Snelgrove, *Continuous-Time Delta-Sigma Modulators for High-Speed A/D Conversion* (Kluwer, Boston, MA, 2000)
3. K. Matsukawa, Y. Mitani, M. Takayama, K. Obata, S. Dosho, A. Matsuzawa, A fifth-order continuous-time delta-sigma modulator with single-opamp resonator. IEEE J. Solid-State Circuits **45**(4), 697–706 (2010)
4. A. Yasuda, H. Tanimoto, T. Iida, A third-order Δ-Σ modulator using second-order noise-shaping dynamic element matching. IEEE J. Solid-State Circuits **33**(12), 1879–1886 (1998)
5. E.N. Aghdam, P. Benabes, Higher order dynamic element matching by shortened tree-structure in delta-sigma modulators. in *Proceedings of European Conference Circuit Theory and Design*, vol. 1, Sept 2005, pp. 201–204
6. R.T. Baird, T.S. Fiez, Improved ΔΣ DAC linearity using data weighted averaging. in *Proceedings of IEEE International Symposium Circuits and Systems (ISCAS)*, vol. 1, May 1995, pp. 13–16

7. W. Yang, W. Schofield, H. Shibata, S. Korrapati, A. Shaikh, N. Abaskharoun, D. Ribner, A 100 mW 10 MHz-BW CT $\Delta\Sigma$ modulator with 87 dB DR and 91dBc IMD. in *IEEE International Solid-State Circuits Conference (ISSCC) Digest of Technical Papers*, Feb 2008, pp. 498–631

8. M.Z. Straayer, M.H. Perrott, A 12-bit, 10-MHz bandwidth, continuous-time $\Sigma\Delta$ ADC with a 5-bit, 950-MS/s VCO-based quantizer. IEEE J. Solid-State Circuits **43**(4), 805–814 (2008)

9. V. Dhanasekaran, M. Gambhir, M.M. Elsayed, E. Sanchez-Sinencio, J. Silva-Martinez, C. Mishra, L. Chen, E. Pankratz, A 20MHz BW 68dB DR CT $\Delta\Sigma$ ADC based on a multi-bit time-domain quantizer and feedback element. in *IEEE International Solid-State Circuits Conference (ISSCC) Digest of Technical Papers*, Feb 2009, pp. 174–175

10. F. Colodro, A. Torralba, New continuous-time multibit sigma-delta modulators with low sensitivity to clock jitter. IEEE Trans. Circuits Syst. I: Regul. Pap. **56**(1), 74–83 (2009)

11. C.-Y. Chen, M.Q. Le, K.Y. Kim, A low power 6-bit flash ADC with reference voltage and common-mode calibration. IEEE J. Solid-State Circuits **44**(4), 1041–1046 (2009)

12. Y.Z. Lin, C.W. Lin, S.J. Chang, A 5-bit 3.2-GS/s flash ADC with a digital offset calibration scheme. IEEE Trans. Very Large Scale Integr. (VLSI) Syst. **18**(3), 509–513 (2010)

13. J. Doernberg, P.R. Gray, D.A. Hodges, A 10-bit 5-Msample/s CMOS two-step flash ADC. IEEE J. Solid-State Circuits **24**(2), 241–249 (1989)

14. B. Verbruggen, J. Craninckx, M. Kuijk, P. Wambacq, G. Van der Plas, A 2.2 mW 1.75 GS/s 5 bit folding flash ADC in 90 nm digital CMOS. IEEE J. Solid-State Circuits **44**(3), 874–882 (2009)

15. S.-W. Chen, R.W. Brodersen, A 6-bit 600-MS/s 5.3-mW asynchronous ADC in 0.13-µm CMOS. IEEE J. Solid-State Circuits **41**(12), 2669–2680 (2006)

16. G. Van der Plas, B. Verbruggen, A 150MS/s 133µW 7b ADC in 90nm digital CMOS using a comparator-based asynchronous binary-search sub-ADC. in *IEEE International Solid-State Circuits Conference (ISSCC) Digest of Technical Papers*, Feb 2008, pp. 242–243, 610

17. J. Craninckx, G. Van der Plas, A 65fJ/conversion-step 0-to-50MS/s 0-to-0.7mW 9b charge-sharing SAR ADC in 90nm digital CMOS. in *IEEE International Solid-State Circuits Conference (ISSCC) Digest of Technical Papers*, Feb 2007, pp. 246–247, 600

18. L. Dorrer, F. Kuttner, P. Greco, P. Torta, T. Hartig, A 3-mW 74-dB SNR 2-MHz continuous-time delta-sigma ADC with a tracking ADC quantizer in 0.13-µm CMOS. IEEE J. Solid-State Circuits **40**(12), 2416–2427 (2005)

19. Y.-C. Lo, H.-P. Chen, J. Silva-Martinez, S. Hoyos, A 1.8V, sub-mW, over 100% locking range, divide-by-3 and 7 complementary-injection-locked 4 GHz frequency divider. in *Proceedings of IEEE Custom Integrated Circuits Conference (CICC)*, Sept 2009, pp. 259–262

20. C.-Y. Lu, Alibrated continuous-time sigma-delta modulators. Ph.D. Dissertation, Department of Electrical and Computer Engineering, Texas A&M University, 2010

21. M. Dessouky, A. Kaiser, Input switch configuration suitable for rail-to-rail operation of switched op amp circuits. Electron. Lett. **35**(1), 8–10 (1999)

22. T. Voo, C. Toumazou, High-speed current mirror resistive compensation technique. Electron. Lett. **31**(4), 248–250 (1995)

23. M. Bazes, Two novel fully complementary self-biased CMOS differential amplifiers. IEEE J. Solid-State Circuits **26**(2), 165–168 (1991)

24. P.E. Allen, D.R. Holberg, *CMOS Analog Circuit Design*, 2nd edn. (Oxford University Press, London, 2002), pp. 477–480

25. G.M. Yin, F.O. Eynde, W. Sansen, A high-speed CMOS comparator with 8-b resolution. IEEE J. Solid-State Circuits **27**(2), 208–211 (1992)

26. L. Sumanen, M. Waltari, V. Hakkarainen, K. Halonen, CMOS dynamic comparators for pipeline A/D converters. in *Proceedings of IEEE International Symposium Circuits and Systems (ISCAS)*, May 2002, vol. 5, pp. V-157–V-160

27. B. Razavi, B.A. Wooley, Design techniques for high-speed, high-resolution comparators. IEEE J. Solid-State Circuits **27**(12), 1916–1926 (1992)

28. P. Amaral, J. Goes, N. Paulino, A. Steiger-Garcao, An improved low-voltage low-power CMOS comparator to be used in high-speed pipeline ADCs. in *Proceedings of IEEE International Symposium Circuits and Systems (ISCAS)*, May 2002, vol. 5, pp. V-141–V-144

29. L.J. Breems, R. Rutten, R. van Veldhoven, G. van der Weide, H. Termeer, A 56mW CT quadrature cascaded ΣΔ modulator with 77dB DR in a near zero-IF 20MHz band. in *IEEE International Solid-State Circuits Conference (ISSCC) Digest of Technical Papers*, Feb 2007, pp. 238–239

30. P. Malla, H. Lakdawala, K. Kornegay, K. Soumyanath, A 28mW spectrum-sensing reconfigurable 20MHz 72dB-SNR 70dB-SNDR DT ΔΣ ADC for 802.11n/WiMAX Receivers. in *IEEE International Solid-State Circuits Conference (ISSCC) Digest of Technical Papers*, Feb 2008, pp. 496–497

31. G. Mitteregger, C. Ebner, S. Mechnig, T. Blon, C. Holuigue, E. Romani, A 20-mW 640-MHz CMOS continuous-time ΣΔ ADC with 20-MHz signal bandwidth, 80-dB dynamic range and 12-bit ENOB. IEEE J. Solid-State Circuits **41**(12), 2641–2649 (2006)

32. M. Park, M. Perrott, A 0.13μm CMOS 78dB SNDR 87mW 20MHz BW CT ΔΣ ADC with VCO-based integrator and quantizer. in *IEEE International Solid-State Circuits Conference (ISSCC) Digest of Technical Papers*, Feb 2009, pp. 170–171, 171a

Chapter 5
An On-Chip Temperature Sensor for the Measurement of RF Power Dissipation and Thermal Gradients

Abstract In this chapter, a design methodology is presented that aims at the extraction of RF circuit performance characteristics from the DC output of an on-chip temperature sensor. Any RF input signal can be applied to excite the circuit under examination because only dissipated power levels are measured, which makes this approach attractive for online thermal monitoring and built-in test scenarios. A fully-differential sensor topology is introduced that has been specifically designed for this method by constructing it with a wide dynamic range, programmable sensitivity to DC and RF power dissipation, as well as compatibility with CMOS technology. Furthermore, a procedure is outlined to model the local electro-thermal coupling between heat sources and the sensor, which is used to define the temperature sensor's specifications as well as to predict the thermal signature of the circuit under test.

5.1 Background

Monitoring performances of individual blocks that constitute a single-chip RF receiver chain is beneficial for identification of faulty devices and self-calibration. In conventional built-in test (BIT) strategies, electrical detectors are placed along the signal path for power measurements [1–4] or extraction of input impedance

This chapter includes portions reprinted with permission, from "Electro-thermal design procedure to observe RF circuit power and linearity characteristics with a homodyne differential temperature sensor," M. Onabajo, J. Altet, E. Aldrete-Vidrio, D. Mateo, and J. Silva-Martinez, accepted for publication in *IEEE Trans. Circuits and Systems I: Regular Papers*, vol. 27, no. 3, pp. 225–240, June 2011, © 2011 IEEE.

matching conditions in the RF front-end [5–7]. Although small, the input impedance of the electrical detectors degrades performance; and the impact of parasitic capacitances from detectors worsens with increasing operating frequencies.

Thermal coupling through the semiconductor substrate generates a rise in temperature in the vicinity of a circuit/device that depends on the device's power dissipation. This thermal coupling can be modeled in the DC domain [8] or with complex small-signal parameters [9]. Moreover, it can be utilized for IC testing purposes [10]. Using on-chip temperature gradients as test observables to measure power dissipation is advantageous because the sensors do not load the circuit under test (CUT) as electrical detectors do. Instead, the small temperature-sensing devices are placed near the CUT, making the technique non-invasive. Furthermore, temperature gradients become more critical to both analog and digital system performance as the integration levels of modern single-chip systems increase, creating incentives to improve diagnosis and compensation techniques. For example, the sensitivity of a direct-conversion receiver in [11] was degraded by 2–4 dB from transient heating effects.

Thermal gradients on a silicon die can be detected with embedded differential temperature sensors [10]. Temperature measurements are usually conducted up to 10 kHz because thermal coupling has low-pass characteristics [9]. But, the multiplication of voltages and currents of different frequencies creates electrical power components at DC and various frequencies [12]. In heterodyne measurement strategies [13], two RF tones at frequencies f_1 and f_2 are applied to the CUT in order to measure the low-frequency power dissipation at $\Delta f = f_2 - f_1$ (<10 kHz) with a temperature sensor. While this approach enables indirect power measurement without interference from on-chip DC temperature gradients, it also necessitates the use of a spectrum analyzer or lock-in amplifier. It is highly desirable to perform measurements at DC to reduce the complexity of the measurement setup and to provide a step towards BIT integration. The RF signal power detected in the thermal DC regime is a result from mixing voltage and current signals at the same frequency, which is why this strategy is referred to as the homodyne method. Since the generated DC temperature gradients are also strongly influenced by the power dissipation in bias circuitry, sensing the RF power requires an on-chip sensor with a wide dynamic range.

This chapter concentrates on a recently developed differential temperature sensor that is feasible for a homodyne BIT strategy [14]. To ensure CMOS compatibility, the sensing devices are formed with parasitic vertical bipolar (PNP) transistors. Section 5.2 provides an overview of this BIT methodology and the application to low-noise amplifier (LNA) characterization is presented as an example. The differential temperature sensor design and tuning features are discussed in Sect. 5.3, for which the measurement results are presented in Sect. 5.4 together with experimental verification of the LNA BIT. Finally, Sect. 5.5 provides conclusions regarding the described concepts and results.

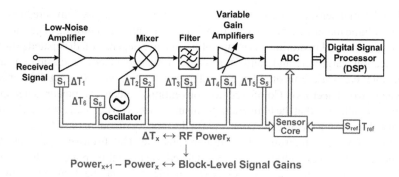

Fig. 5.1 Generalized receiver diagram with on-chip thermal sensing

5.2 Temperature Sensing Approach

5.2.1 Integration with Transceiver Calibration Techniques

A temperature sensing strategy is appealing for BIT applications where the goal is to: (1) identify gross failures that affect the power dissipation in bias circuitry; (2) measure the signal power along processing paths; (3) design self-calibration schemes that can adapt to temporary thermal hot spots occurring near a sensitive circuit. The envisioned purpose of a homodyne sensing scheme is illustrated in Fig. 5.1, in which several small temperature-sensing devices (S_i, where i ranges from 1 to 6 in Fig. 5.1) are located at various test points within analog blocks of an RF receiver and at one reference location (S_{ref}). In a system-on-a-chip, the temperature gradients between the sensing devices S_i and S_j ($i \neq j$) or S_i and S_{ref} can be acquired through processing the sensor core output signals. This larger sensor core contains the necessary bias and amplification circuits to provide a DC output to an on-chip analog-to-digital converter (ADC). If the on-chip ADC is not available for reuse, then a dedicated 8–12 bit low-power (<50 μW) ADC with 0.05–0.7 mm² die area would be sufficient for online digitization of the DC sensor output at a low sampling rate (e.g. 100 kHz as in [15, 16]). In such a case, the total area overhead of the sensor core, 20 sensing devices, the ADC, and 0.75 mm² room for reference voltage and bias current generation circuitry would be between 2 and 15% for a 10–25 mm² receiver chip. Finally, the comparisons of the differential measurements conclude in the digital signal processor (DSP), allowing DC temperature gradients and the signal power (i.e. gain) along the analog receiver chain to be monitored. As a step towards realizing such a system-level BIT, the focus in this chapter is on the measurement of the RF power dissipation and 1 dB compression point of an LNA. In brief, the goal is to illuminate design considerations for a practical sensor circuit that can be employed as on-chip detector near analog blocks for system-level calibration methods as those described in Sect. 2.2. Another potential use of the sensors is to monitor the average power dissipating in digital blocks for the detection of faults.

The thermal sensing approach could be used for low-cost pass/fail screening in a high-volume manufacturing test environment or for online monitoring of parameter drift during normal operation. As temperature linearly depends on dissipated power, specification variations and faults that cause a change in power dissipation are detectable; e.g. variations of either S_{11} in the front-end or gains (output differences between two detectors). Compared to conventional electrical power detectors, a major advantage is that the temperature-sensing devices do not load the signal paths because they are not electrically connected to the input or output of the CUTs, leaving only the coupling path through the common substrate. The discussed approach can also be extended to an individual die in a stacked-die assembly, but each die should include its own reference sensing device (S_{ref} in Fig. 5.1) and the differential power gain comparisons should only be made for test points on the same die because each die has its own common-mode temperature.

5.2.2 Modeling of the Thermal Coupling

Various modeling [8–10, 17, 18] and simulation strategies [19, 20] exist to account for the static and dynamic effects of thermal coupling on the performance of electrical devices on the same die. In this BIT application, the primary interest lies in estimating the temperature increase from power dissipation in the CUT at the location of the sensing device. Hence, the silicon substrate has been modeled with an RC network in order to allow coupled analysis with the electrical behavior of the CUT and temperature sensor using the Spectre simulator in Cadence.

Figure 5.2 displays the RC network for the example layout scenario described throughout this chapter. The three-dimensional silicon die has been modeled with 5 layers in the vertical z-direction. Each node in the RC network models a unit volume of the die whose dimensions can be selected based on the trade-off between accuracy and simulation time. Here, a cube size ($x_u \times y_u \times z_u$) of $10 \times 10 \times 10$ µm was chosen for the surface (1st) layer to approximate the distances between points. This grid size was selected because it is comparable with sensing device dimensions, which implies that the devices are approximated as having the same size as the unit grid. With this model, the electrical voltage at each node in the network is equivalent to a temperature change in degrees (Kelvin or Celsius) relative to the ambient die temperature during the electro-thermal co-simulation, and any injected electrical current is equivalent to power dissipation of a device located at the node. Capacitors in the network can be omitted if only DC temperature analysis is needed, but they are included in this example to predict settling times and to maintain a generic model that accounts for frequency-dependence. Points M, C, and R in Fig. 5.2 represent the locations of devices (Table 5.1) from which dissipated power (in Watts) is injected into the network modeled as current (in Amperes). As shown in Fig. 5.3, these current sources are connected to the equivalent points M, C, and R in Fig. 5.2 based on the layout locations of the devices. The local temperature change is measured with a parasitic

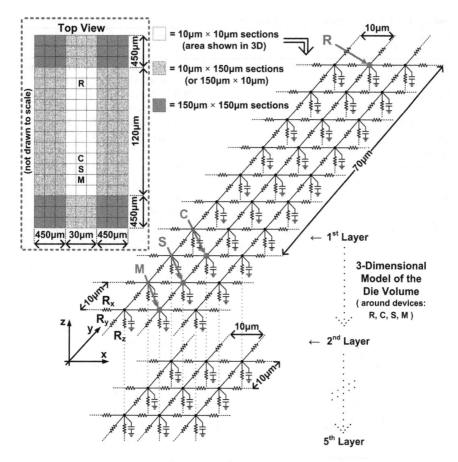

Fig. 5.2 RC network model for electro-thermal coupling. The parameters are based on distances between point heat sources (M, C, R) and a sensing device (S) in the actual chip layout

vertical PNP device at point S having spacing in the layout of 7 and 10 μm from points C and M, respectively. The temperature (T_s) change of the temperature transducer in the sensor is obtained by coupling the voltage at node S to the PNP device through an ideal voltage-controlled voltage source with gain of k = −1.8 mV/K to modulate the base-emitter voltage (V_{be}) of the PNP transistor according to its temperature sensitivity [21]. Here, the temperature-dependence is assumed to be linear over the range of interest.

In the RC network model (Fig. 5.2), the less critical layers 2 through 5 have z-direction lengths of 20, 40, 80, and 160 μm to model 310 of the 330 μm thick substrate. To reduce the complexity and simulation time, the fine-resolution grid (shown in 3D) was only extended by 10 μm around points M, S, C, and R; while low-resolution unit volumes with the following dimensions were employed at the sides and corners to expand the grid by 450 μm into the horizontal directions (only shown in the top view): 10 × 150 μm × z_u, 150 × 10 μm × z_u, and 150 × 150 μm × z_u.

Table 5.1 CUT design parameters and simulation results

Component/ Specification	Dimensions/Value at 1 GHz
M_M (point M)	W/L = 7.2 μm × 13 fingers/0.18 μm (layout area: 12 × 37 μm)
M_C (point C)	W/L = 7.2 μm × 25 fingers/0.18 μm (layout area: 11 × 41 μm)
R_L (point R)	100 Ω (layout area: 22 × 35 μm)
Technology/V_{DD}	0.18 μm CMOS/2.4 V
I_{DC}	8.7 mA
Gain (S_{21})	0.8 dB[a]
1 dB Compression Point	0.5 dBm
S_{11}	−11.7 dB
S_{22}	−10.6 dB

[a] The LNA is loaded (without buffer) by an additional external 50 Ω impedance from measurement equipment and additionally by the estimated packaging/PCB parasitics

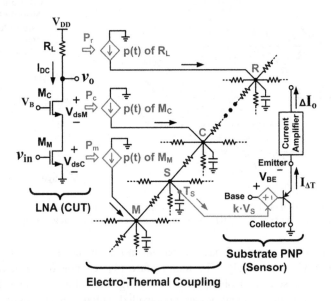

Fig. 5.3 Electro-thermal coupling between CUT and sensing device

Finally, the lateral edges are terminated with infinite impedances and the bottom of the 5th layer is grounded, i.e. the thermal boundary conditions are assumed adiabatic and isothermal, respectively. Each discretized capacitance and the directional node resistances in Fig. 5.2 are calculated as follows [11]:

$$C = \rho \cdot c \cdot x_u \, y_u \, z_u, \tag{5.1}$$

$$R_x = x_u / (\kappa \cdot y_u \, z_u), \tag{5.2}$$

$$R_y = y_u/(\kappa \cdot x_u \, z_u), \tag{5.3}$$

$$R_z = z_u/(\kappa \cdot x_u \, y_u) \tag{5.4}$$

where the mass density (ρ), specific heat capacity (c), and thermal conductivity for silicon (κ) are $2.3 \cdot 10^6 \text{g/m}^3$, $0.7 \text{ J/(g} \cdot \text{K)}$, and $120 \text{ W/(m} \cdot \text{K)}$ at 75°C, respectively [11].

5.2.3 Electro-Thermal Analysis Example: Low-Noise Amplifier

Figure 5.3 depicts the main devices of the CUT, the PNP sensing device, and how the RC network couples both circuits. The CUT is a typical broadband LNA with resistive load for which design details can be found in Table 5.1 and [22]. Next, it will be shown that circuit-level power and linearity characteristics of blocks can be extracted using temperature sensors even with a single test tone. However, in system-level testing strategies, multi-tone tests or a frequency sweep of a single test tone typically enhance the fault coverage. Assuming a sinusoidal signal with voltage amplitude A at v_{in} in Fig. 5.3 and combining the DC analysis with the small-signal analysis, simplified expressions for the average power dissipation of the devices can be derived in terms of the transconductances (g_{mM}, g_{mC}) and DC drain-source voltages (V_{dsM}, V_{dsC}) of the transistors (M_M, M_C), load resistor R_L, and DC current I_{DC}:

$$R_L : \quad P_r = R_L \cdot I_{DC}^2 + \frac{1}{2}(g_{mM} \cdot A)^2/R_L, \tag{5.5}$$

$$M_M : \quad P_m = V_{dsM} \cdot I_{DC} - \frac{1}{2}(g_{mM} \cdot A)^2/g_{mC} \tag{5.6}$$

$$M_C : \quad P_C = V_{dsC} \cdot I_{DC} - \frac{1}{2}(g_{mM} \cdot A)^2 \cdot (R_L - 1/g_{mC}) \tag{5.7}$$

Here, the energy conservation principle holds since the AC amplitude-dependent terms sum up to zero and $P_r + P_m + P_c = V_{DD} \cdot I_{DC}$. The above expressions show that the average power from the RF signal adds to the DC power at the load resistor but subtracts from the DC power at the active devices acting as RF power sources. This property implies that the ideal placement of the temperature-sensing PNP device in the layout is either on the side of the load resistor that does not face the MOS transistors, or between the two transistors where their temperature effects add. The latter location was selected as shown in Fig. 5.4. Resistor R_L was placed more than 50 μm away from the sensor to reduce thermal interference, which can be assessed by injecting the power of R_L at a point R on the RC network in Fig. 5.3 during the simulations.

The broadband LNA used as the CUT in Fig. 5.3 was designed with 11 dB gain for on-die probing [22]. Table 5.1 lists the key design and performance parameters

Fig. 5.4 Area of the die with
CUT (LNA) and temperature-
sensing PNP device

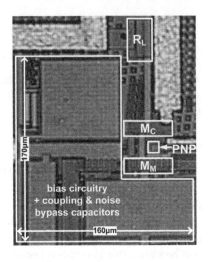

from simulations of this LNA with estimated parasitics for the packaged prototype chip. The graphs in Fig. 5.5 were obtained by sweeping the RF power of a single-tone input to the CUT and plotting the average power for each device. As expected from (5.5) to (5.7), the DC component of the dissipated power due to RF signal processing adds to the DC bias power at the resistor and subtracts from the DC bias power at the MOS transistors. The analysis in Appendix D explains how the nonlinearities of the MOS transistors cause their DC power curves (P_m, P_c) to have minima. Notice that the DC component of the power due to RF circuit activity is significantly less than the DC bias power dissipation of each device, which translates into a high dynamic range requirement when the same sensor should be capable to measure the effects of DC bias as well as of RF signal processing via temperature changes. In addition, the sensor must at least have sufficient sensitivity to detect a change in the dissipated DC power from 20 to 200 μW associated with the −10 to 0 dBm electrical signal input power levels.

Figure 5.6 visualizes the simulated local temperature change T_s at node S shown in Figs. 5.2 and 5.3. The DC bias of the CUT creates static 0.996°C change of T_s with respect to the ambient temperature. As the amplitude of the electrical signal applied to the CUT input increases, the local temperature changes as a result of the superimposed thermal coupling from the power dissipations (Fig. 5.5) in devices M_M, M_C, and R_L. The DC power/temperature reaches a minimum that can be related to the 1 dB compression point with a shift on the x-axis (Appendix D). The simulation result in Fig. 5.6 also indicates that the sensor sensitivity should be high enough to detect 5–30 m°C changes in the −15 to 0 dBm range of interest. The CUT and electro-thermal network were simulated with −5 dBm input power to assess the transient response of the temperature change. Figure 5.7 reveals that the settling time is approximately 8 μs, which is adequately short for production testing.

Fig. 5.5 Simulated average powers at devices in the CUT vs. RF input power. *Top* P_r at R_L, *middle* P_m at M_M, *bottom* P_c at M_C

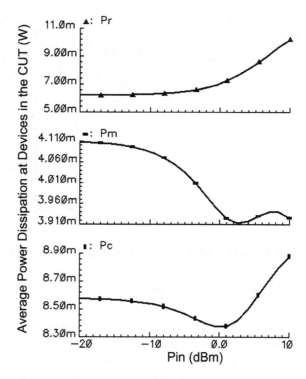

Fig. 5.6 Temperature change T_s at the sensing device vs. RF input power

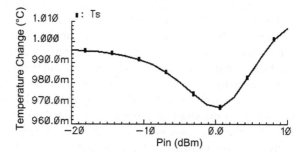

Fig. 5.7 Transient behavior of T_s with -5 dBm input power

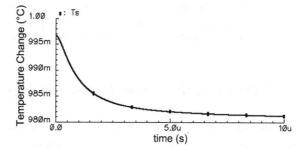

Fig. 5.8 A differential
CMOS temperature sensor
with lateral PNP devices.
(This circuit was proposed in
[23].)

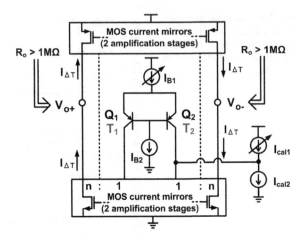

5.3 CMOS Differential Temperature Sensor Design

5.3.1 Previous Sensors

Various passive and active sensors for on-chip differential temperature measurements are experimentally compared in [23], and a schematic representation of a previously presented CMOS-compatible fully-differential sensor is shown in Fig. 5.8. Conceptually, the two temperature-sensing parasitic PNP devices (Q_1, Q_2) are placed as a differential pair within an operational transconductance amplifier (OTA) configuration. The collector current difference between Q_1 and Q_2 due to temperature difference $\Delta T = T_1 - T_2$ is amplified by current mirrors within the OTA before flowing into the high impedance nodes at the output. Currents I_{cal1}/I_{cal2} can be adjusted to compensate for electrical and thermal offsets. This sensor has a high sensitivity of up to ~ 400 mV/mW when the CUT that dissipates power is placed at 20 µm distance from Q_1 (or Q_2) and there is a spacing of 400 µm between Q_1 and Q_2. A drawback of this topology is its limited dynamic range of less than 1.5 mW with this sensitivity. Generally, such differential sensors with high sensitivity are optimal for the heterodyne approach [12, 13] and the AC setup at low frequencies in [23] with external lock-in amplifier or spectrum analyzer. Since the heterodyne measurements of two RF tones at the Δf frequency are free of interference from DC temperature gradients, the previous sensors are well-suited to sense and amplify the low-power mixing product at Δf without saturating the sensor.

5.3.2 Design of the Sensor Circuit Topology

In this book, the focus is on the homodyne measurement approach and the development of a sensor core optimized for application to RF BIT measurements at DC without relying on any external equipment. Hence, the sensor must have a

Fig. 5.9 Wide dynamic range differential temperature sensor. (The devices Q_1 and Q_2 are vertical parasitic PNP transistors in a CMOS process.)

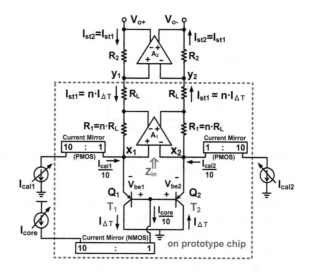

wide dynamic range to enable concurrent DC and RF power measurements. Additionally, differential temperature sensors are often comprised of lateral parasitic PNP devices, but some CMOS processes only model vertical PNP devices which are more restrictive because the collector (p-type substrate) is typically grounded. Parasitic vertical PNP devices are popular temperature sensors because they offer high precision and repeatability; e.g. $\pm 0.1°C$ absolute error from -50 to $130°C$ in [24], where the error can be treated as DC offset and V_{be} temperature sensitivity spread due to process variations is limited to below 2% depending on the technology.

Figure 5.9 displays the sensor topology that was constructed with vertical PNP devices [14]. Sensing transistors Q_1 and Q_2 are biased with the same operating point, having common base and collector terminals. These two devices can be either the sensing or reference points (S_i or S_{ref}) in Fig. 5.1. The DC emitter voltages are also forced to be identical due to the virtual ground created by the feedback from the first amplifier (A_1). Note that the collector current difference of Q_1 and Q_2 under this DC bias ideally only depends on the temperature difference ($\Delta T = T_1 - T_2$) between their respective locations. In practice, device mismatches and thermal gradients cause offsets that can be compensated with currents I_{cal1} and I_{cal2}. The temperature-dependent differential current ($I_{\Delta T}$) is amplified with a cascade of a transimpedance amplifier (TIA) stage (A_1, R_1) and resistive load $R_L = R_1/n$ connected to a virtual ground from a subsequent TIA stage (A_2, R_2). Consequently, the current amplification ($I_{st1} = n \cdot I_{\Delta T}$) depends on reliable resistive matching to minimize sensitivity variations. Moreover, the sensitivity can be changed with the base current I_{core} [10] to allow reuse of the same sensor near low- and high-power devices on the chip and to compensate for any process-dependent gain variations. As a proof-of-concept, the sensor core and first amplification stage with R_L were implemented on the prototype chip, while stage 2 was realized with an off-chip amplifier for simplified external DC voltage

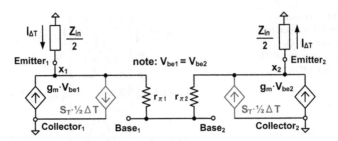

Fig. 5.10 Simplified small-signal equivalent circuit of the sensor core. (The PNP devices are represented with the hybrid-π model.)

measurements. In a BIT application, the output current of stage 1 could be digitized directly or the second amplification stage could be included on the chip.

The dynamic range improvement with the discussed sensor topology comes from the virtual ground at nodes $x_{1,2}$ in Fig. 5.9, which furnishes a low impedance at the emitters of $Q_{1,2}$. It also avoids that $I_{\Delta T}$ is converted to a voltage difference at the emitters, which would cause imbalance of the bias conditions of the sensing PNP devices. Instead, the current is processed by a low-gain TIA stage with a controlled amplification ratio.

A simplified small-signal model of the PNP pair in the sensor core is shown in Fig. 5.10. The temperature difference causes a change of the emitter current with a sensitivity that can be roughly estimated as $S_T \approx k \cdot g_{mQ}$; where k and g_{mQ} are the temperature sensitivity of the base-emitter voltage and the transconductance of $Q_{1,2}$, assuming $Z_{in} \ll 1/(k \cdot g_{mQ})$. Part of this temperature-dependent current will not be amplified by the TIA because it will flow through resistance r_{π}, resulting in unavoidable sensitivity loss. It is important to minimize the effective load impedances at the emitters presented by the input impedance (Z_{in}) of the first TIA stage. A high amplifier gain improves the overall sensitivity by lowering Z_{in} in Fig. 5.9 according to the following approximation:

$$Z_{in} \approx R_1 / \left[1 + A_1 \left(\frac{R_L || R_1}{R_L || R_1 + r_o} \right) \right] = \frac{R_1}{1 + A_v} \qquad (5.8)$$

where r_o and A_v are the output resistance and loaded voltage gain of amplifier A_1. To determine the appropriate gain prior to the design of amplifier A_1, the sensor core was simulated with an ideal amplifier model having a variable gain. The sensitivity ($\Delta I_{st1}/\Delta T$) vs. A_v is plotted in Fig. 5.11 and a target value of $A_v \approx 32 \approx 30$ dB was selected from these simulations to avoid major efficiency degradation in the sensor core. Additionally, matched polysilicon resistors of $R_1 = 8$ kΩ and $R_L = 1$ kΩ were selected for a robust current amplification ratio of $n = 8$. To ease testing of this prototype design, R_2 was an off-chip 100 kΩ resistor and A_2 was an off-the-shelf operational amplifier (NJM4580D) with 110 dB DC gain.

Figure 5.12 shows the schematic of amplifier A_1. It consists of a simple differential pair (M_2) loaded by transistors (M_3) in saturation region and a PMOS source follower output stage (M_4, M_5). The amplifier's input DC level depends on the bias conditions

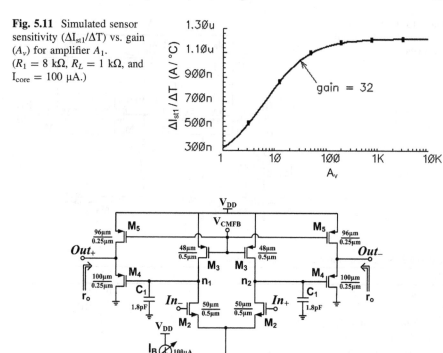

Fig. 5.11 Simulated sensor sensitivity ($\Delta I_{st1}/\Delta T$) vs. gain (A_v) for amplifier A_1. ($R_1 = 8$ kΩ, $R_L = 1$ kΩ, and $I_{core} = 100$ μA.)

Fig. 5.12 Amplifier (A_1) schematic with annotated width/length dimensions

of $Q_{1,2}$ in the sensor core (Fig. 5.9), which is why nodes $n_{1,2}$ are regulated by the common-mode feedback (CMFB) circuit in Fig. 5.13. M_5 in the source-follower stage is also connected to the output of the CMFB circuit, and the regulated voltage level at $n_{1,2}$ is transferred to the output nodes through the gate-source voltage drop across M_4, resulting in an output DC level around 1.55 V. A PMOS source-follower stage was selected over an NMOS stage to increase the voltage headroom in the sensor core by allowing more voltage drop across R_1 in Fig. 5.9. Since only DC amplification is required, capacitors (C_1) were included at the internal high-impedance nodes to create gain roll-off that approximates a single-pole response to stabilize the amplifier. Its simulated performance with CMFB is summarized in Table 5.2.

5.3.3 Adjustment of the Sensor's Sensitivity

DC simulations of the standalone sensor circuit can be performed by sweeping the SPICE parameter *Trise* of one PNP device to emulate its temperature increase above the ambient temperature due to local heating from the CUT. For example,

Fig. 5.13 Common-mode
feedback (CMFB) circuit
schematic

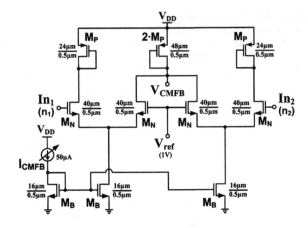

Table 5.2 Simulated amplifier (A_1) specifications

Parameter	Value
DC Gain	30.2 dB
f_{3dB}	1.74 MHz
Unity gain frequency (f_u)	56.9 MHz
Phase margin	89.7°
Integrated input-referred noise (DC $-$ f_u)	55.1 μV
Common-mode rejection ratio[a]	75.5 dB at 10 kHz
Power supply rejection ratio[a]	36.4 dB at 10 kHz
Output resistance	270 Ω
5% Settling time (1 mV step input, unloaded)	264 ns
CMFB loop: DC Gain/Phase margin	35.1 dB/74.4°
Input offset voltage (standard deviation)	1.5 mV
Technology/V_{DD}	0.18 μm CMOS/1.8 V
Power dissipation (with CMFB)	1.05 mW

[a] For a single output. The fully-differential processing in the sensor topology improves the noise
rejection

the plots in Fig. 5.14 were generated this way in order to evaluate the dynamic
range based on the output current I_{st1} of the first amplification stage in Fig. 5.9.
The results show that the linear range is ±4.7°C with 2.94 μA/°C sensitivity and
±13.4°C with 0.99 μA/°C sensitivity for $I_{core} = 1$ mA and $I_{core} = 100$ μA,
respectively. The sensor core's wide dynamic range with adjustable sensitivity is
sufficient to monitor devices with power beyond 50 mW. Large differential output
currents cause a large voltage drop across R_1 in Fig. 5.9, which forces $M_{4,5}$ in the
amplifier (Fig. 5.12) out of the saturation region.

Currents I_{cal1} and I_{cal2} (Fig. 5.9) permit the compensation of DC temperature
gradients as well as electrical offsets from mismatches in the cascaded amplifier
stages. The appropriate calibration current ranges can be determined with DC
simulations that include anticipated electrical device mismatches while modeling
the heat sources of the CUT or any other nearby circuits in the simulation based on

Fig. 5.14 Simulated
dynamic range of the sensor
core

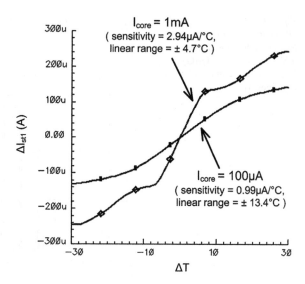

Fig. 5.3. For example, $I_{cal1} = 100$ μA compensates for an equivalent thermal offset at the sensing device location of approximately 8°C (0.99 μA/°C sensitivity setting). Offset voltages are also calibrated out. Based on the Monte Carlo simulation results in Fig. 5.15, the V_{be} mismatch of the PNP pair and the input offset of amplifier A_1 have standard deviations of 0.8 and 1.5 mV, respectively; and the simulated V_{be} mismatch due to absolute temperature changes from −50 to 130°C is less than 0.2 mV (Fig. 5.16). In the calibration step preceding a measurement, the sensor can be balanced by adjusting I_{cal1} and I_{cal2} under monitoring of the differential output until it is close to 0 V. This was done manually in the experimental characterization (Sect. 5.4.1), but could be performed with the same on-chip ADC that resolves the sensor output in a system-level BIT scenario (Fig. 5.1).

5.3.4 Sensor Design Optimization Procedure

To perform co-simulations of the CUT and appropriate sensor circuit it is advisable to follow these steps:

(1) Construct the electro-thermal coupling network described in Sect. 5.2.2 based on the actual or anticipated layout locations of the devices in the CUT. The capacitors can be removed if only DC analysis is to be performed.
(2) Select a suitable layout location to place a single parasitic PNP transistor near the device(s) to be monitored, and perform the simulation in Sect. 5.2.3 which will reveal the temperature change at the related node in the grid. Select a suitable location for the reference parasitic transistor that will be used to process the thermal gradient. In this example, Q_2 is located at a distance of

Fig. 5.15 Assessment of offsets in the sensor core with Monte Carlo simulations. **a** V_{be} mismatch of Q_1/Q_2, **b** input offset voltage of amplifier A_1

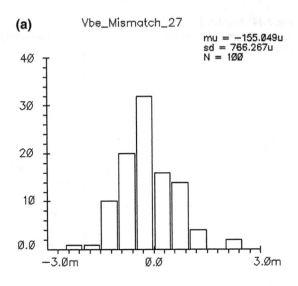

(a)

Vbe_Mismatch_27

mu = −155.049u
sd = 766.267u
N = 100

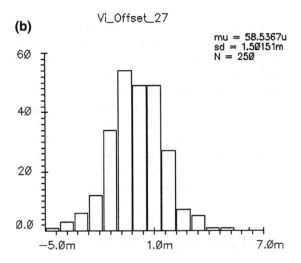

(b)

Vi_Offset_27

mu = 58.5367u
sd = 1.50151m
N = 250

420 µm where the simulated temperature change is about two orders of magnitude lower than at the sensing device Q_1.

(3) Determine the required dynamic range and temperature sensitivity for the sensor from the results in (2). In the previously discussed example, average power dissipations of 4.1 and 8.6 mW at M_M and M_C caused almost a 1°C imbalance between the PNP transistors. A wide dynamic range is desirable to monitor low- and high-power devices on a chip. On the other hand, a sensitivity around 5 m°C is needed to detect the RF signal power at the LNA. Hence, the sensor circuitry must have sufficient gain to achieve this resolution. Notice that, if this technique is utilized to characterize other blocks, then the

Fig. 5.16 Simulated V_{be} mismatch of Q_1/Q_2 vs. ambient temperature

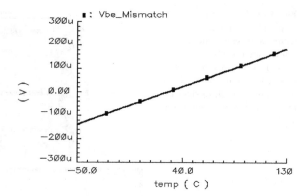

Fig. 5.17 Combined CUT and sensor simulation. The plot shows the differential sensor output voltage after settling vs. average RF input power applied to the CUT having a 1 dB compression point of 0.5 dBm

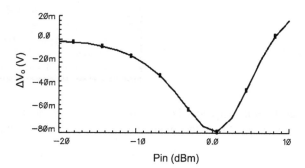

higher power levels of the signals processed in the receiver chain makes it easier to sense the temperature changes.

(4) Design a differential temperature sensor circuit consisting of the PNP transistor pair in step (2) as well as bias and amplification circuitry to meet the specifications in step (3). Nodes in the extended RC network allow assessing that the temperature change at the reference PNP device (Q_2) is significantly smaller than at Q_1 near the CUT. In the presented case, the DC temperature changes are 0.996°C at Q_1 and 96, 20, 13 m°C at 150, 300, 450 μm away from Q_1 respectively. In an integrated system, effects from circuits further than 150 μm away from the CUT are attenuated by more than one order of magnitude, but their impacts can be accounted for by injecting their power dissipations as currents into the extended RC grid.

(5) Simulate the CUT, electro-thermal network, and complete sensor circuit by coupling the schematics as in Fig. 5.3. Optimize the sensing device placement as well as the sensor circuit's gain, dynamic range, and transient response based on the simulated electro-thermal coupling.

As example, the plot in Fig. 5.17 was obtained with a CUT/sensor co-simulation to assess the 0.5dBm 1 dB compression point identification capability, showing that the sensor output reaches a −79.0 mV minimum with 0.63 dBm input power. Based on the

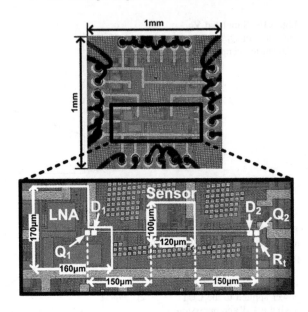

Fig. 5.18 Micrograph of the chip with differential temperature sensor and LNA. Emitter area of $Q_{1,2}$: 11×11 μm. Area of sensor core: 0.012 mm^2 (reusable with additional Q_x devices to monitor multiple locations on a die)

analysis in Appendix D, the simulated relative input power shift (0.13 dB) should be subtracted from the minimum power point to predict the 1 dB compression point.

5.4 Measurement Results

Figure 5.18 displays the microphotograph of the chip fabricated in Jazz Semiconductor 0.18 μm 1P6 M CMOS technology. Sensing device Q_2 (11×11 μm) is located at a reference point that is separated from active devices of the sensor core by 150 μm. Additional diode-connected MOS transistors ($D_{1,2}$ with W/L = 60/0.18 μm) and a 50 Ω polysilicon resistor R_t (5×33.8 μm) are placed 4 μm away from the sensing devices as extra test heat sources. Standard multimeters were used for the measurement of voltage drops and currents to determine the DC power at these heat sources.

5.4.1 Temperature Sensor Characterization

Figure 5.19 shows the measured differential output voltage in response to the DC power dissipation at resistor R_t, which was kept below 16 mW to prevent damage based on the process–specific recommendations for the device and interconnect dimensions. The plots show that the linear range with 199.6 mV/mW sensitivity is slightly above 12 mW, but it extends beyond 16 mW in the 41.7 mV/mW sensitivity setting. Although 16 mW dynamic range is adequate to monitor

Fig. 5.19 Sensor output vs. power dissipation at resistor R_t. The measurements were performed with $I_{core} = 100$ µA (sensitivity = 41.7 mV/mW) and $I_{core} = 1$ mA (sensitivity = 199.6 mV/mW). Distance between R_t and Q_2: 4 µm

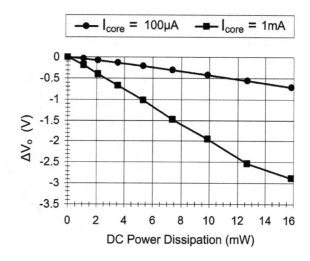

Fig. 5.20 Sensor output vs. power of diode-connected MOS transistors $D_{1,2}$. The measurements were performed with $I_{core} = 100$ µA (sensitivity = 42.0 mV/mW) and $I_{core} = 1$ mA (sensitivity = 207.9 mV/mW). Distance between $D_{1,2}$ and $Q_{1,2}$: 4 µm

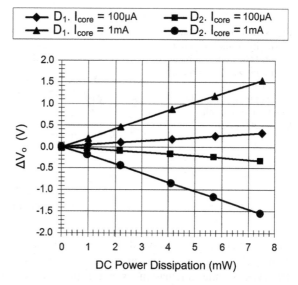

conventional high-power devices, the simulations (Fig. 5.14) indicate that the range is more than 30 mW with $I_{core} = 100$ µA (41.7 mV/mW sensitivity).

Figure 5.20 displays the plots from the sensor characterization measurements in which the DC power in the diode-connected transistors $D_{1,2}$ near each sensing device (Q_1, Q_2) was swept individually up to the safe limits for the particular device layouts. The results reveal the symmetric nature of the fully-differential circuitry and that the sensitivity to power in the MOS devices is approximately the same as for the resistor within the sensor's linear range, which can also be observed from the sensitivity vs. I_{core} plots in Fig. 5.21.

To verify that the sensor has a high rejection to ambient temperature changes, D_1 and D_2 were excited concurrently by injecting DC currents and adjusting the

Fig. 5.21 Sensitivity control to power in R_t and $D_{1,2}$ via I_{core} adjustments

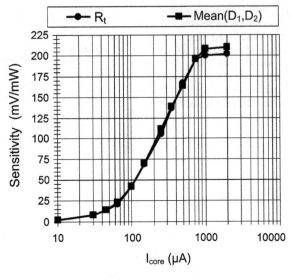

Fig. 5.22 Common-mode sensitivity of the temperature sensor. (The common-mode sensitivity was measured by sweeping the power dissipation in D_1 and D_2 simultaneously with $I_{core} = 500$ μA.)

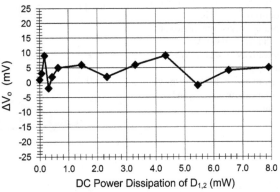

currents such that the measured DC power in both devices is identical for each data point in Fig. 5.22. Even though the sensitivity to common-mode power is below 10 mV/mW, the fluctuations suggest that the sensor calibration step should precede the CUT measurement if the ambient temperature is expected to have changed significantly since the last measurement.

The offset calibration range was evaluated under three conditions: (1) when the test heat sources (D_1, D_2, R_t) do not dissipate power and with a deactivated LNA (named "Heat OFF"), (2) with an activated LNA and 3.9 mW additional power dissipation in D_1 (named "Heat ON") to achieve $\Delta V_o \approx 0$ V with $I_{cal1} = I_{cal2} = 0$, (3) when R_t alone dissipates 15.9 mW. Case (1) gives insight into the ability to recover from the sensor's inherent electrical offsets due to component mismatches without interference from the LNA's DC bias. As shown in Fig. 5.23, the differential output voltage has a linear dependence on the calibration currents as long as the electrical amplification stages in the sensor are

Fig. 5.23 Offset calibration
with currents I_{cal1} and I_{cal2}
($I_{core} = 500$ µA)

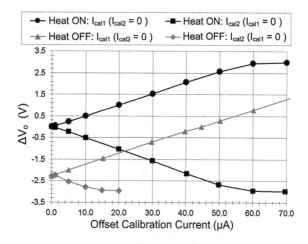

Fig. 5.24 Offset calibration
range with I_{cal1} ($I_{cal2} = 0$,
$I_{core} = 500$ µA)

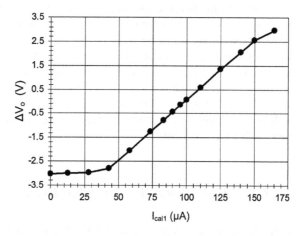

not saturated, and $I_{cal1} = 44.6$ µA is required to compensate for on-chip and
off-chip component variations of this prototype design. Case (2) makes it evident
that heat sources can also be used to balance the sensor, which in this case requires
3.9 mW power in D_1 in addition to the DC bias power of the LNA to achieve
$\Delta V_o \approx 0$ V. Furthermore, the plot in Fig. 5.23 under the "Heat ON" condition
shows the symmetry of the output voltage dependence on I_{cal1} and I_{cal2}. In case (3),
the 15.9 mW power dissipation at R_t without activation of other heat sources
creates an extreme imbalance in the operating conditions of the two bipolar
transistors due to both the offset from process variations and the extra temperature
gradient. The measured sensor output voltage for this case is plotted versus I_{cal1} in
Fig. 5.24, demonstrating that $I_{cal1} = 95.6$ µA establishes a balanced output and
that the offset compensation capability spans the linear range of the sensor
circuitry. The offset calibration currents were adjusted to compensate for DC
temperature gradients and electrical offsets by obtaining $\Delta V_o \approx 0$ V prior to each
set of measurements under certain bias conditions, which requires adjustments in

Table 5.3 Measured CUT[a] performance parameters

Parameter	Value at 1 GHz
Gain (S_{21})	−2.3 dB[b]
1 dB Compression point	0.5 dBm
Third-order intercept point (IIP3)	12.0 dBm
S_{11}	−6.3 dB
S_{22}	−12.7 dB
I_{DC}	8.7 mA
Technology/V_{DD}	0.18 μm CMOS/2.4 V

[a] LNA loaded (without buffer) by a 50 Ω analyzer impedance
[b] Reduced due to the external 50 Ω load in addition to the on-chip load resistor (R_L) and due to S_{11} degradation from packaging/PCB parasitics at 1 GHz; $S_{21} \approx 0$ dB up to 500 MHz

the micro-ampere range. In practice, the ADC and digital post-processing will limit the test time because the settling times of the temperature change (Fig. 5.7) and amplifier (Table 5.2) are below 10 μs and 500 ns, respectively. However, up to 18 clock cycles could be required for the calibration phase (assuming 6-bit programmability for the calibration test sources and a binary search algorithm until $\Delta V_o \approx 0$ V), averaging of several sensor output measurements (might be required in a noisy system-on-chip environment), and test control operations. At a 100 KS/s rate, this would imply 0.18 ms per test point. The test time could be even shorter with the availability of a faster on-chip ADC or off-chip test resources in a production test environment.

5.4.2 RF Testing with the On-Chip DC Temperature Sensor

Table 5.3 gives an overview of the CUT parameters that are relevant to the correlation of its RF output and the temperature sensor output. The RF measurements were taken around 1 GHz because the parasitics of the QFN package and PCB assembly degraded S_{11} to worse than −6.3 dB at higher frequencies. Losses from cables, power combiner, bias-T, and impedance mismatches were characterized and de-embedded from the measurements reported below. A spectrum analyzer was used to measure the CUT output while simultaneously reading the differential sensor output with a DC voltmeter in order to experimentally verify that the CUT's RF performance can be extracted with temperature sensor measurements. To correlate measurements with simulations, Fig. 5.25 contains plots of the CUT and sensor outputs from a sweep of the RF input power applied to the CUT with a single tone at 1 GHz. Offsets on the y-axes are caused by the ~3 dB CUT gain difference between simulations and measurements with extra losses. The curves show that input power levels above −15 dBm can be monitored at the output of this DC sensor, which is sufficient when a signal with more power than a typical LNA input signal is applied during testing. Online

Fig. 5.25 Measurement vs. simulation comparison for the CUT characterization. The plots show the LNA's 1 dB compression point curve and the DC output voltage of the sensor with $I_{core} = 500$ µA (167 mV/mW sensitivity)

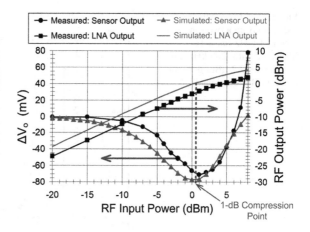

Fig. 5.26 LNA output power and log-magnitude of the sensor output voltage. I_{core} was 500 µA (167 mV/mW sensitivity) during these measurements

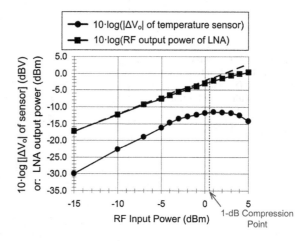

testing with input signals below -15 dBm would require sensor sensitivity improvements. Options that can be explored are designing the sensor with more amplification or implementing $Q_{1,2}$ with PNP devices that are electrically connected in Darlington configuration to boost the gain and to increase the coupling to the CUT surrounded by two nearby PNP devices (Q_1).

In Fig. 5.25, the minimum of the temperature sensor's ΔV_o curve is -71 mV with 1dBm input power. Subtracting the fixed 0.13 dB shift according to the simulations results in Sect. 5.3.4, the estimated 1 dB compression point is 0.87 dBm. This value approximates the electrically measured 1 dB compression point with an error of 0.37 dB, which is comparable to standard RF power detectors in BIT applications. As described in Appendix D, estimation inaccuracies create further uncertainty of ± 0.6 dB, yielding up to 1 dB error for the 1 dB compression point prediction.

Compared to the simulated plot in Fig. 5.25, it can be observed that the measured minimum is about 10% higher due to electro-thermal modeling

Fig. 5.27 CUT's output spectrum from a two-tone test around 1 GHz (case 1). Measured with: 200 kHz tone spacing, −22.2 dBm per tone (−19.2 dBm combined)

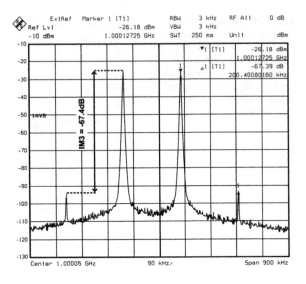

Fig. 5.28 CUT's output spectrum from a two-tone test around 1 GHz (case 2). Measured with: 200 kHz tone spacing, −2.2 dBm per tone (1.2 dBm combined)

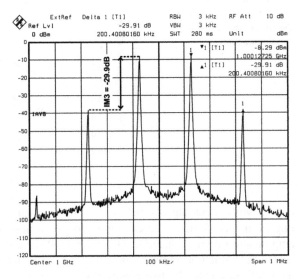

inaccuracies. This discrepancy is acceptable since the sensitivity of the sensor can be adjusted with I_{core} over a tuning range of roughly a decade (Fig. 5.21). A log-magnitude plot of the measured sensor output voltage vs. CUT input power is displayed in Fig. 5.26 to visualize how the 1 dB compression point corresponds to the vicinity of the peak log-magnitude of the sensor output voltage.

Figure 5.27 displays the CUT's output spectrum around 1 GHz that was obtained with two −22.2 dBm test tones having a separation of 200 kHz. As reference, the third-order intermodulation (IM3) of −67.4 dB is annotated for this linear operating condition. The 1 dBm input power level was identified as critical nonlinear point based on the temperature sensor output measurements.

For comparison, Fig. 5.28 shows the output spectrum with two −2.2 dBm test tones that have a combined power of 1.2 dBm. The resulting IM3 is −29.9 dB, which demonstrates the usefulness of this point as indicator for nonlinear operation. Since the DC temperature sensor characterization of the CUT circumvents the use of RF measurement equipment, it provides a viable alternative to monitor RF signal levels and linearity performance in BIT applications and pass/fail production testing in which a 1 dB error is permissible.

5.5 Summarizing Remarks

A sensing methodology was described that exploits the intrinsic down-conversion of circuit performance information from the RF domain to the DC domain with the homodyne temperature measurement approach. It was shown that this property is useful for application in built-in testing and monitoring of on-chip thermal gradients that can impact system performance. Since this alternative technique does not require a connection to the circuit under test or the signal path, it provides a non-influential method for monitoring variations. The presented CMOS-compatible sensor architecture has been developed for the wide dynamic range and programmability requirements as built-in power detector based on the homodyne approach.

Furthermore, an electro-thermal design procedure for differential temperature sensors was experimentally validated. Coupling at low frequencies could impact the CUT's operation, which can be evaluated with electro-thermal simulations. Measurement results obtained with an RF amplifier and a 0.012 mm^2 built-in temperature sensor on a 0.18 μm CMOS test chip revealed that the same sensor can detect the DC and RF power dissipation, and that the 1 dB compression point can be predicted from the sensor's output with an error below 1 dB without RF measurement equipment.

References

1. Q. Yin, W.R. Eisenstadt, R.M. Fox, T. Zhang, A translinear RMS detector for embedded test of RF ICs. IEEE Trans. Instrum. Meas. **54**(5), 1708–1714 (2005)
2. S. Bhattacharya, A. Chatterjee, Use of embedded sensors for built-in-test RF circuits. in *Proceedings of IEEE International Test Conference (ITC)*, Oct 2004, pp. 801–809
3. Q. Wang, M. Soma, RF front-end system gain and linearity built-in test. in *Proceedings of 24th IEEE VLSI Test Symposium*, May 2006, pp. 228–233
4. A. Valdes-Garcia, R. Venkatasubramanian, J. Silva-Martinez, E. Sánchez-Sinencio, A broadband CMOS amplitude detector for on-chip RF measurements. IEEE Trans. Instrum. Meas. **57**(7), 1470–1477 (2008)
5. J.-Y. Ryu, B.C. Kim, I. Sylla, A new low-cost RF built-in self-test measurement for system-on-chip transceivers. IEEE Trans. Instrum. Meas. **55**(2), 381–388 (2006)

6. T. Das, A. Gopalan, C. Washburn, P.R. Mukund, Self-calibration of input-match in RF front-end circuitry. IEEE Trans. Circuits Syst II: Express Briefs **52**(12), 821–825 (2005)
7. X. Fan, M. Onabajo, F.O. Fernández-Rodríguez, J. Silva-Martinez, E. Sánchez-Sinencio, A current injection built-in test technique for RF low-noise amplifiers. IEEE Trans. Circuits Syst. I: Regul. Pap. **55**(7), 1794–1804 (2008)
8. D.J. Walkey, T.S. Smy, R.G. Dickson, J.S. Brodsky, D.T. Zweidinger, R.M. Fox, Equivalent circuit modeling of static substrate thermal coupling using VCVS representation. IEEE J. Solid-State Circuits **37**(9), 1198–1205 (2002)
9. N. Nenadovic, S. Mijalkovic, L.K. Nanver, L.K.J. Vandamme, V. d'Alessandro, H. Schellevis, J.W. Slotboom, Extraction and modeling of self-heating and mutual thermal coupling impedance of bipolar transistors. IEEE J. Solid-State Circuits **39**(10), 1764–1772 (2004)
10. J. Altet, A. Rubio, E. Schaub, S. Dilahire, W. Claeys, Thermal coupling in integrated circuits: application to thermal testing. IEEE J. Solid-State Circuits **36**(1), 81–91 (2001)
11. S. Mattisson, H. Hagberg, P. Andreani, Sensitivity degradation in a tri-band GSM BiCMOS direct-conversion receiver caused by transient substrate heating. IEEE J. Solid-State Circuits **43**(2), 486–496 (2008)
12. D. Mateo, J. Altet, E. Aldrete-Vidrio, J. L. Gonzalez, Frequency characterization of a 2.4 GHz CMOS LNA by thermal measurements. in *Proceedings of IEEE Radio Frequency Integrated Circuits (RFIC) Symposium*, June 2006, pp. 565–568
13. J. Altet, E. Aldrete-Vidrio, D. Mateo, X. Perpiñà, X. Jordà, M. Vellvehi, J. Millán, A. Salhi, S. Grauby, W. Claeys, S. Dilhaire, A heterodyne method for the thermal observation of the electrical behavior of high-frequency integrated circuits. Meas. Sci. Technol. **19**(11), pp. 115704 (8 pp), Nov 2008
14. M. Onabajo, J. Altet, E. Aldrete-Vidrio, D. Mateo, J. Silva-Martinez, Electro-thermal design procedure to observe RF circuit power and linearity characteristics with a homodyne differential temperature sensordifferential temperature sensor. IEEE Trans. Circuits Syst. I: Regul. Pap. **58**(3), 458–469 (2011)
15. M.D. Scott, B.E. Boser, K.S.J. Pister, An ultralow-energy ADC for smart dust. IEEE J. Solid-State Circuits **38**(7), 1123–1129 (2003)
16. N. Verma, A.P. Chandrakasan, An ultra low energy 12-bit rate-resolution scalable SAR ADC for wireless sensor nodes. IEEE J. Solid-State Circuits **42**(6), 1196–1205 (2007)
17. L. Codecasa, D. D'Amore, P. Maffezzoni, Modeling the thermal response of semiconductor devices through equivalent electrical networks. IEEE Trans. Circuits Syst. I: Fundam. Theor. App. **49**(8), 1187–1197 (2002)
18. V. Szekely, On the representation of infinite-length distributed RC one-ports. IEEE Trans. Circuits Syst. **38**(7), 711–719 (1991)
19. S.-S. Lee, D.J. Allstot, Electrothermal simulations of integrated circuits. IEEE J. Solid-State Circuits **28**(12), 1283–1293 (1993)
20. W. Van Petegem, B. Geeraerts, W. Sansen, B. Graindourze, Electrothermal simulation and design of integrated circuits. IEEE J. Solid-State Circuits **29**(2), 143–146 (1994)
21. J. Michejda, S.K. Kim, A precision CMOS bandgap reference. IEEE J. Solid-State Circuits **19**(6), 1014–1021 (1984)
22. H.M. Geddada, J.W. Park, J. Silva-Martinez, Robust derivative superposition method for linearising broadband LNAs. Electron. Lett. **45**(9), 435–436 (2009)
23. E. Aldrete-Vidrio, D. Mateo, J. Altet, Differential temperature sensors fully compatible with a 0.35 μm CMOS process. IEEE Trans. Compon. Packag. Technol. **30**(4), 618–626 (2007)
24. M.A.P. Pertijs, G.C.M. Meijer, J.H. Huijsing, Precision temperature measurement using CMOS substrate pnp transistors. IEEE Sens. J. **4**(3), 294–300 (2004)

Chapter 6
Mismatch Reduction for Transistors in High-Frequency Differential Analog Signal Paths

Abstract An analog calibration technique is described in this chapter, which aims at lessening the mismatch between transistors in the differential high-frequency signal path of analog CMOS circuits. It can be applied for offset reduction in high-speed amplifiers and comparators in which short-channel devices are utilized to minimize bandwidth reduction from parasitic capacitances. In general, it is suitable for RF applications in which direct matching of the transistors is undesired because sophisticated layout practices would increase the coupling between high-frequency signal paths. The methodology involves auxiliary devices that sense the existing mismatch as part of a feedback loop for error minimization. This mismatch reduction technique is demonstrated for a differential amplifier as well as for IIP2 improvement of a double-balanced mixer.

6.1 Background

Until now, the approaches discussed in this book are mostly aimed at making analog and mixed-signal circuits more robust by either circumventing their dependence on mismatches or by introducing digitally programmable elements for post-fabrication adjustments. An alternative approach to deal with rising variability is to decrease the mismatches of analog circuits by lessening them in a statistical sense. The approach discussed in this chapter is targeting the static mismatch between critical transistors in particular, where the goal is to decrease the standard deviation of the parameter variations by employing an automatic analog calibration loop.

Device mismatches become more severe as technology scaling continues, especially when minimum transistor dimensions are used to optimize for high-speed operation or to bias with high overdrive voltage for yield enhancement [1]. In addition to higher percent errors for small fabrication dimensions, the threshold

M. Onabajo and J. Silva-Martinez, *Analog Circuit Design for Process Variation-Resilient Systems-on-a-Chip*, DOI: 10.1007/978-1-4614-2296-9_6, © Springer Science+Business Media New York 2012

voltage mismatch worsens even for neighboring transistors due to the increasing effect of dopant fluctuations in modern CMOS processes [2]. The resulting offsets degrade the performance of analog circuits that rely on device matching. For example, the second-order intermodulation intercept point (IIP2) of mixers strongly depends on matching of transistors, for which a digital mismatch reduction scheme was proposed in [3] to adjust gate bias voltages separately for each switching transistor.

Another issue in RF circuit design is that designers might place transistors next to each other with a safe distance instead of elaborately matching them in the layout. Even though the use of non-minimum dimensions can reduce process variations, devices with large area (i. e., large parasitic capacitances) in the signal path are often not feasible since they imply increased power consumption and/or performance degradation, which is the case in high-speed amplifiers and comparators [4]. Similarly, layout matching techniques such as interleaved or common-centroid styles create more high-frequency coupling through parasitic capacitances of crossing metal lines or leakage through the substrate due to the proximity of the devices. An alternative design technique towards the goal of alleviating transistor mismatches is described in this chapter. The method involves an analog calibration loop in which device mismatches are indirectly detected and reduced through layout-based parameter correlations rather than directly measuring characteristics of the circuit. This calibration loop continuously operates in the background without requiring digital resources or switches in the signal path. Its short convergence time below 10 μs prevents excessive start-up calibration time for time-critical situations such as during production testing.

6.2 A Mismatch Reduction Technique for Differential Pair Transistors

6.2.1 Approach

In RF applications, designers may choose to place transistors next to each other with a safe distance as shown in Fig. 6.1 instead of matching them in the layout. The advantage with such a configuration is that the physical separation of the devices provides isolation against RF signal leakage that leads to crosstalk between the differential signal paths. Often, each RF transistor is surrounded by a guard ring for enhanced isolation and by deep trenches (if available). A drawback in this scenario is that the unmatched devices have significant parameter mismatches which are observable through the static drain current difference.

To alleviate the mismatch problem, the alternative approach visualized in Fig. 6.2 is presented here. Instead of matching the RF transistors M_1 and M_2 to each other, they are individually matched to mismatch-sensing transistors M_{1S} and M_{2S} in a DC calibration loop. Thus, the currents I_{1S} and I_{2S} of the mismatch-sensing transistors are correlated to I_1 and I_2 of the main transistor pair, respectively. Even though it is

Fig. 6.1 An unmatched
RF transistor pair

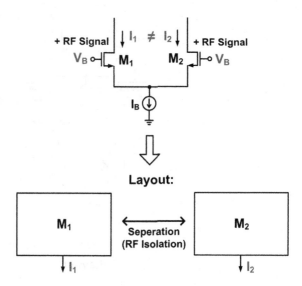

optimal to use the same dimensions and number of fingers for M_{1S} and M_{2S} as for M_1 and M_2, they do not have to be identical. However, their electrical device parameters must be correlated to M_1 and M_2 through layout matching techniques. The feedback action in the loop compares I_{1S} to I_{2S} and adjusts the separate gate bias voltages V_{B1} and V_{B2} of the mismatch-sensing transistors until the currents are approximately equal to each other. Consequently, the drain current difference in the main transistor pair is also reduced due to the parameter correlations between the matched transistors and the shared gate bias voltages. In this way, the mismatches are lessened while the RF isolation between the main transistors is maintained. Additionally, low-pass filter nodes within the calibration loop suppress any RF signal that might couple into it through layout parasitics.

To demonstrate the aforementioned concept, Fig. 6.3 depicts a differential amplifier consisting of a transistor pair (M_1, M_2) with polysilicon resistor loads (R_L), where the resistor dimensions were selected large enough to ensure that the input-referred offset voltage is dominated by M_1 and M_2. Table 6.1 lists the device dimensions for the circuit. The characteristics of M_1 and M_2 are ideally equal, but considerable deviations occur when they are not matched in the layout through interleaved, common-centroid, or similar configurations. Hence, crosstalk between the differential signal paths is avoided by physically separating them, while parameter variations of M_1 and M_2 should be treated as uncorrelated. However, M_1 and M_2 can be laid out with N (=20 in this example) subdevices, and matched to sensing-transistors M_{1S} and M_{2S} respectively. In this configuration, M_{1S} and M_{2S} are part of the DC calibration loop that detects a mismatch between currents I_1 and I_2, and that generates bias voltages V_{B1} and V_{B2} individually for each branch. If the drain currents of M_{1S} and M_{2S} are forced to be equal in the absence of mismatches, then their gate-source voltage overdrives must be equal [2], which only occurs when $V_{C1} = V_{C2}$ in Fig. 6.3. Here, M_{1S} and M_{2S} are placed in a differential amplifier

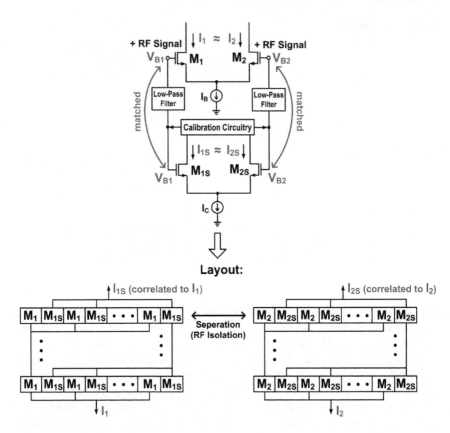

Fig. 6.2 An RF transistor pair with DC mismatch reduction loop

configuration with a tail-current source ($I_B/10$) and active loads (M_3, M_4) for high gain with self-regulation via feedback resistors (R_{cm}). Capacitors (C_{st}) stabilize the loop by creating a dominant pole at nodes V_{C1} and V_{C2}. If $I_1 \neq I_2$ in the presence of device mismatches, then the resulting imbalance of $V_{C1} - V_{C2}$ is amplified by the amplifier (A). The feedback action differentially adjusts V_{B1} and V_{B2} until $V_{C1} \approx V_{C2}$ to minimize mismatches without requiring on-chip digital resources. Capacitors (C_{filt}) are included to filter out high-frequency noise. Amplifier A, whose schematic is shown in Fig. 6.4, controls the bias voltages V_{B1} and V_{B2} around a set common-mode output level ($V_B = 0.85$ V). Its transistor dimensions in the nominal corner case (Table 6.1) were selected according to this required DC level, and its feedback resistors (R_{fb}) provide regulation in the presence of device mismatches.

This scheme exploits that the parameters of M_1/M_{1S} (and M_2/M_{2S}) are highly correlated so that the mismatch can be continuously extracted in the background to compensate for drifts from temperature changes as well as process variations. Since the calibration loop has several low-pass filtering nodes, the differential signal integrity is not jeopardized by coupling between M_1 and M_2 through the loop. Instead, coupling to M_{1S}/M_{2S} via layout parasitics and substrate leakage due

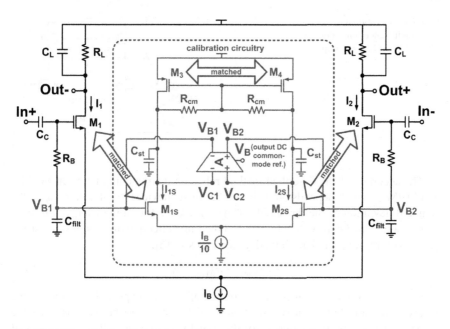

Fig. 6.3 Differential amplifier with transistor mismatch reduction loop

Table 6.1 Differential
amplifier and calibration loop
components

Component	Dimensions/Value
M_1, M_2, M_{1S}, M_{2S}	W/L = 90 nm × 20 fingers/90 nm
M_3, M_4	W/L = 6.25 μm × 8 fingers/3.7 μm
R_L	1.12 kΩ (L/W = 9/2 μm)
C_L	0.1 pF
R_B	100 kΩ
C_{filt}	1 pF
C_c	5 pF
C_{st}	10 pF
R_{cm}	100 kΩ (L/W = 20 × 10/1 μm)
I_B	1 mA
Technology/Supply Voltage	90 nm CMOS/1.2 V
Operational Transconductance Amplifier (A):	
M_N	W/L = 8 μm × 4 fingers/4 μm
M_P	W/L = 5 μm × 2 fingers/1.55 μm
M_B	W/L = 3 μm × 4 fingers/1 μm
R_{fb}	38 kΩ (L/W = 8 × 19/1 μm)
I_{bias}	50μA
DC Gain: Amplifier, Calibration Loop	18.3 dB, 38.6 dB

to the matching only create small signal losses. The large bias resistors (R_B) prevent that the input capacitances looking into the gates of M_{1S} and M_{2S} cause any significant loading effects at the RF inputs (In+, In−).

Fig. 6.4 Operational
transconductance amplifier
(A) in the calibration loop

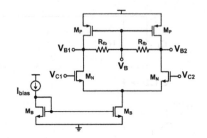

The accuracy of the described method relies on the matching between M_1/M_{1S} and M_2/M_{2S}, which depends on their number of subdevices [5, 6]. Let $\sigma_{\Delta Vth}$ be the standard deviation of the threshold voltage difference for an unmatched transistor pair. In a matched pair with N fingers and the same effective dimensions in a stripe pair structure, this standard deviation decreases to [6]:

$$\sigma_{\Delta Vth(m)} = \sigma_{\Delta Vth}/\sqrt{N} \qquad (6.1)$$

More complex common-centroid configurations are expected to improve the spread reduction, but the abovementioned relationship will be used as plausible worst-case estimate. Being outside of the signal path, the parasitic capacitances of the matched transistors in the core of the calibration loop do not affect the RF performance. Hence, their dimensions can be increased to ensure that their offsets are negligible. Therefore, non-minimum transistor lengths (L) and widths (W) were selected (Table 6.1, Figs. 6.3, 6.4) for matched pairs M_3/M_4, M_B, M_N, M_P, I_B (NMOS current mirror), and $I_B/10$ based on the inverse proportionality of $\sigma_{\Delta Vth}$ to $\sqrt{(W \cdot L)}$ [7]. Likewise, the polysilicon resistors R_{cm} and R_{fb} were sized sufficiently large with the help of statistical device models and Monte Carlo simulations.

6.2.2 Simulation Results

The test circuit (Figs. 6.3, 6.4) was designed using UMC 90 nm CMOS technology with a 1.2 V supply, and simulations were performed with the foundry's statistical device models. The loaded differential amplifier under calibration has a gain of 13 dB with a -3 dB bandwidth of 2.14 GHz. Its minimum AC input impedance magnitude within the passband is 1.77 kΩ, which changes less than 1% when the loop is added and activated. The amplifier (A) has a loaded DC gain of 18.3 dB, resulting in an overall gain of 38.6 dB in the loop starting at V_{B1}/V_{B2} and traversing through V_{C1}/V_{C2}.

Device matching was taken into account during the Monte Carlo analysis with the Cadence Spectre simulator (process and mismatch variations enabled) by calculating the expected spread reduction in Eq. 6.1 based on the number of fingers in each matched pair. According to this reduction, the corresponding correlation coefficient (C_m) was specified from the relation given in [8]:

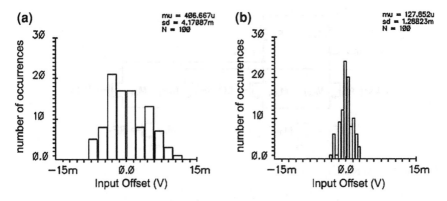

Fig. 6.5 Monte Carlo simulation results (100 runs at 30°C). Input-referred offset voltage of the differential amplifier: **a** without mismatch calibration (ideal bias voltages: $V_{B1} = V_{B2} = V_B$), **b** with activated mismatch calibration loop

$$1/\sqrt{N} = \sqrt{1 - C_m} \qquad (6.2)$$

For example, the $N = 20$ fingers ($C_m = 0.95$) of the matched pairs M_1/M_{1S} and M_2/M_{2S} leads to an expected spread reduction of 4.47 with this scheme when other offsets in the loop are negligible. Figure 6.5 displays the histograms of the input-referred offset voltage of the amplifier obtained with 100 Monte Carlo runs at 30°C, showing that its standard deviation decreases from 4.17 to 1.29 mV when the calibration loop is added. Notice that an input offset decrease from 4.17 to 1.29 mV corresponds to a drain current difference reduction from 3.1% to 1.0% for M_1 and M_2). At −40°C and 100°C, 100 Monte Carlo runs revealed that the predicted offset decreases from 4.10 to 1.22 mV and from 4.25 to 1.40 mV, respectively. With the large-sized devices in the calibration circuit, the accuracy improvement mainly depends on the correlation of the parameters between M_1/M_{1S} (and M_2/M_{2S}). For instance, using $C_m = 0.99$ in the simulation instead of the previous worst-case assumption, the input offset with calibration reduces to 0.76 mV (0.6% M_1/M_2 drain current difference); provided that 20 subdevices can be appropriately matched with a common-centroid layout.

6.3 Second-Order Nonlinearity Enhancement for Double-Balanced Mixers

6.3.1 Introduction

Second-order nonlinearity of the down-conversion mixer is typically the bottleneck for the overall achievable second-order intermodulation intercept point (IIP2) performance with direct-conversion and low-IF receiver architectures [9, 10],

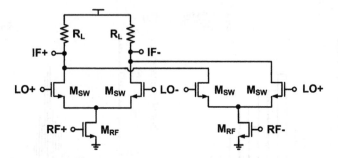

Fig. 6.6 Double-balanced mixer

which are appealing low-power architectures for low-cost portable wireless devices. Thus, stringent IIP2 specification demands are imposed on mixers in these systems, especially with the tendency towards wider bandwidths that leads to increased interference signals at the RF front-end. For instance, a minimum mixer IIP2 requirement of 60 dBm has been identified for the UMTS receiver design budget in [11]. Similarly, the down-conversion mixer IIP2 for WCDMA systems has been specified as 59 dBm in [12], whereas the IIP2 target for the WCDMA/CDMA2000 mixer in [13] was 50 dBm. Even though the IIP2 mixer requirement depends on the given communication standard and system-level design, 50 dBm can be regarded as the minimum tolerable mixer IIP2 for direct-conversion receivers based on findings in the literature. A general approach to derive this mixer specification is given in [14], where even the need for mixers with IIP2 > 70 dBm has been outlined.

6.3.1.1 IIP2 Degradation Mechanisms with Ideal Switching Transistors in the Core

The schematic of a double-balanced mixer [15] is displayed in Fig. 6.6, in which the bias circuitry is omitted for simplicity. Transistors labeled M_{RF} are the input transconductors to which the RF signal is applied. Assuming a hard-switching local oscillator (LO) signal and the corresponding square-wave approximation, it has been shown in [9] that the IIP2 can be estimated with the following equation when the switching core transistors (M_{SW}) are considered ideal:

$$IIP2 \approx \frac{\sqrt{2}}{\pi \eta_{nom} \alpha_2} \times \frac{4}{2 \cdot \Delta\eta(\Delta g_m + \Delta A_{RF}) + \Delta R_L(1 + \Delta g_m)(1 + \Delta A_{RF})} \quad (6.3)$$

Parameters g_m, α_2, and Δg_m in (6.3) are the nominal transconductance, second-order non-linearity coefficient, and transconductance deviation of the two transistors M_{RF}. ΔA_{RF} is the amplitude difference at the RF+ and RF− inputs, and ΔR_L is the discrepancy between the two load resistors. The nominal LO duty cycle is η_{nom}, which has an associated mismatch of $\Delta\eta$ between LO+ and LO−. It is

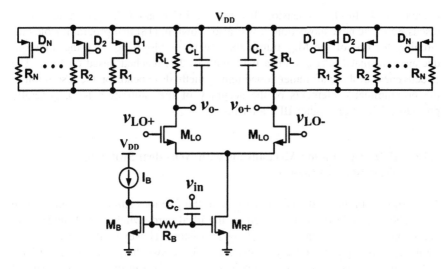

Fig. 6.7 IIP2 tuning with digital load resistance trimming

worthwhile to point out that the $\Delta\eta$ term exclusively depends on the LO signal under the ideal switching core assumption, but it becomes strongly affected by threshold voltage offsets of the switching transistors in the practical case. As discussed later in this chapter, this switching transistor-dependent IIP2 degradation can be as severe as the degradation due to load mismatches.

It can be observed from Eq. 6.3 that any mismatches between the branches deteriorate the IIP2. Furthermore, the adverse effects from Δg_m and ΔA_{RF} scale with ΔR_L and $\Delta\eta$, implying that the fundamental IIP2 limit depends primarily on the load resistor and LO signal/transistor mismatches. The second term in the second denominator gives rise to the importance of accurate load resistor matching [9]. For this reason, adjustable loads consisting of parallel resistors with switches were proposed in [16] to compensate for process variations as shown in Fig. 6.7. The measurement results of this work have shown that 5-bit programmability in the mixer load resistors leads to receiver IIP2 improvements in the 23–26 dB range. Analogously, when the mixer load contains current sources, an additional feedback loop can be added to reduce the common-mode output impedance mismatch for approximately 20 dB IIP2 enhancement [17].

Revisiting Eq. 6.3, another observation is that any of the parameters in the second denominator can be tuned to minimize this mismatch-dependent denominator. Various IIP2 improvement schemes involve tuning of parameters other than the load mismatch. In [18] for example, an LO buffer with tunable phase for one of the differential outputs was used to change the duty cycle term ($\Delta\eta$) in order to maximize IIP2. Alternatively, LO duty cycle modification is also possible by adjusting the gate bias voltages of the individual LO transistors, which affects the turn on/off time instants of the switches [19]. However, notice that such an approach will impact the maximum achievable IIP2 limit under consideration of mismatches in the LO transistors because the LO bias conditions are altered as discussed in the next subsection. It was also shown in [18] that programmable bias circuitry for one of the M_{RF} transistors can be employed to vary the transconductance mismatch (Δg_m) until a maximum IIP2 is reached based on (6.3). The effectiveness of these aforementioned tuning methods depends on the resolution of the programmable elements or the accuracy of the calibration loop, generally providing 20–30 dB higher IIP2 after tuning.

6.3.1.2 IIP2 Degradation Mechanisms with Non-Ideal Switching Transistors in the Core

The results with the methods summarized in the previous subsection demonstrate the capabilities of IIP2 tuning based on the ideal hard-switching LO model with negligible mismatches in the switching transistors. However, the intrinsic IIP2 limit depends primarily on the mismatches in the switching transistors [10] for fully-differential double-balanced mixers (e.g., the mixer in Fig. 6.6 with a shared tail current source added at the sources of the M_{RF} transistors). In the pseudo-differential case (e.g., the mixer in Fig. 6.6 without any modifications), the intrinsic IIP2 limit depends predominantly on the input transconductor as well as on the switching transistor mismatch, where a common-mode feedback circuit at the IF output can be used to suppress the input transconductor's contribution to the IIP2 [20]. This makes the mismatch of the LO switching transistors critical for the achievable best-case IIP2.

Let L be the low-frequency leakage parameter due to mismatches between the M_{SW} transistors in Fig. 6.6. A detailed expression for L can be found in [10], but it is important to point out here that this parameter is zero for perfectly matched M_{SW} transistors, and that it is directly proportional to the relative offset voltages of non-ideal M_{SW} transistors. Thus, L is a statistically-varying mismatch parameter. Its impact on the RMS voltage of the IIP2 is evident from the following equation [10]:

$$\sigma_{IIP2} = \frac{(2/\pi) \cdot g_m}{\sqrt{L^2 \cdot [(\alpha_2^{dif})^2 (\alpha_2^{cm})^2] + [(\Delta R_L / R_L) \cdot \alpha_2^{cm}]^2}} \qquad (6.4)$$

where R_L and ΔR_L are the load resistors in Fig. 6.6 and their mismatch, respectively. As before, g_m is the transconductance of the RF input transistor M_{RF}, whose second-order nonlinearity has a differential component α_2^{dif} and a common-mode component α_2^{cm}. Equation 6.4 reflects that load resistor mismatch only degrades IIP2 in the presence of α_2^{cm}, which is alleviated when a common-mode feedback is added [20] or when fully-differential input transconductors with high common-mode rejection at low frequencies (within IF bandwidth) are employed [11]. On the other hand, the mismatches of the LO switching transistors limit the achievable IIP2 through parameter L and the combined input transconductor nonlinearities. Even if α_2^{cm} is made negligible by designing with high common-mode rejection, the differential second-order nonlinearity α_2^{dif} will deteriorate IIP2 with non-perfectly matched LO transistors. The approach presented in [3] aims at canceling the offset between the LO transistors by using separate digitally programmable gate bias voltages. With regards to Eq. 6.4, this means a reduction of parameter L by setting the switches to the exact combination that gives minimal offsets between the transistors, resulting in simulated (theoretical) IIP2 improvements up to roughly 40 dB with 6-bit resolution of the bias adjustment voltage. The mixer calibration technique in Sect. 6.3.2 applies the automatic analog calibration scheme from Sect. 6.2 for reduction of the LO transistor mismatches in order to boost the intrinsic IIP2 limit based on Eq. 6.4.

6.3.1.3 IIP2 Calibration with Digital Control

Regardless of which mechanisms degrade the IIP2, a DC offset can be dynamically injected at the output of the mixer to improve the IIP2. A system-level IIP2 calibration technique has been demonstrated in [21] by injecting an offset current at the mixer output with a digitally controllable current source having 6-bit resolution. Such a scheme is aligned with the system-level calibration approach discussed in Sect. 2.2.4. The ADC output in the receiver is analyzed in the digital signal processor to control the offset current sources based on the digitally measured static and dynamic DC offsets. Similarly, the calibration in [22] involves an auxiliary second-order intermodulation (IM2) generator that cancels the IM2 in the mixer. The IM2 generator contains a programmable scaling unit that can be adjusted for optimum IIP2 performance when IIP2 monitoring capabilities exist on the chip. Another digital calibration method utilizes a least-mean-square (LMS) algorithm operating on the digitized output of a common-mode detector at the mixer output and the baseband filter's output to tune the IIP2 by injecting a DC current [23]. Even though digital approaches are effective and allow calibration control through the DSP, they typically involve significantly longer convergence times compared to analog control loops. Additionally, they rely on DSP resources for the measurement of performance degradation and the corresponding corrective actions, which might not be available on the chip with the RF front-end circuitry.

6.3.1.4 Autonomous IIP2 Reduction/Cancellation

The benefits and trade-offs of digital and analog circuit-level calibrations have been discussed in the subsections of Sect. 2.2. Instead of using digitally programmable elements to tune IIP2, automatic analog feedback loops can be employed as well. The work in [24] is a representative paradigm for analog IIP2 calibration, which involves an IM2 generator whose output determines how much current is injected into the mixer core to cancel the IM2. With such a scheme, the amount of IIP2 improvement (e.g., 22 dB from simulations in [24]) depends on the gain in the feedback loop. In theory, the IM2 component with calibration is given by

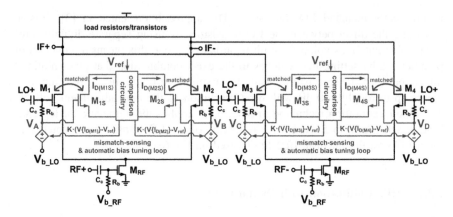

Fig. 6.8 Mixer with conceptual mismatch reduction for the LO transistors

$$IM2_{cal} = \frac{IM2_i}{1 + A_L} \tag{6.5}$$

where $IM2_i$ is the IM2 without calibration and A_L is the loop gain. In practice, the calibration circuitry must be designed with care to avoid that component offsets and mismatches degrade its effectiveness. Since the calibration loop bandwidth is typically in the range of the IF signal bandwidth, the required frequency response is usually achievable using non-minimum device dimensions to lessen mismatches.

Another reported IIP2 improvement method involves cancellation of the input transconductor's second-order nonlinearity parameter α_2 in Eq. 6.3 with a modified bias network that serves as IM2 generator [25]. Simulations of this alternative approach indicate that 20–40 dB IIP2 improvement is achievable with this method even though it does not involve a feedback loop.

6.3.2 Alternative Mixer Calibration

In the following case study, the objective is to improve the intrinsic IIP2 of a double-balanced down-conversion mixer by reducing the mismatches of the LO switching transistors that proportionally increase the leakage parameter L in Eq. 6.4. This method is intended for applications in which limited on-chip digital computational resources are available or in which a fast analog IIP2 tuning at start-up helps to reduce the convergence time and required range of a digital system-level calibration algorithm.

Figure 6.8 gives an overview of the calibration for a double-balanced mixer based on the mismatch reduction loop discussed in Sect. 6.2. Here, the goal is to force equal currents in the calibration branches ($I_{D(M1S)} \approx I_{D(M2S)} \approx I_{D(M3S)} \approx I_{D(M4S)}$),

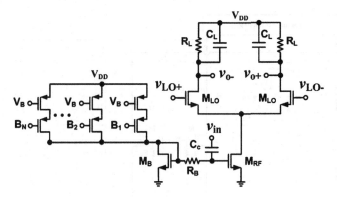

Fig. 6.9 Digital DC bias control for mixer gain tuning

minimizing their mismatches and the corresponding mismatches in the transistors of the mixer that are switched by the LO signal.

The comparison circuitry in Fig. 6.8 utilizes the same mechanism to accomplish the mismatch reduction as the calibration loop described in Sect. 6.2. In this circuit, the LO transistors M_1–M_4 are assumed to be matched to the associated mismatch-sensing transistors M_{1S}–M_{4S} in the layout, which results in the parameter correlations described in Sect. 6.2. Within the comparison circuitry, all currents from the sensing transistors are converted to a voltage $V\{I_{D(Mx)}\}$ which is then compared to a common reference V_{ref}. The difference is amplified by a factor K within the control loops for the individual bias voltages V_A–V_D. These gate bias voltages are shared by each LO transistor and its mismatch-sensing transistor, and they are controlled around the gate bias voltage V_{b_LO} with which the mixer is designed. Notice that bias resistors (R_b) and coupling capacitors (C_c) form high-pass filters that allow the RF signals to pass, whereas the DC mismatch calibration circuitry contains low-pass filters (not shown). The high-valued resistors (R_b) further isolate the calibration circuitry from the LO signal. It is also worth mentioning that the gate bias voltage V_{b_RF} for the input transconductor in Fig. 6.8 is independent of the calibration loop and available for tuning. In receivers with I/Q paths, this gate bias voltage of the transconductor M_{RF} can be adjusted for I/Q amplitude matching of the mixer outputs in both paths [16], which is visualized in Fig. 6.9.

The key building blocks of the calibration scheme are displayed in Fig. 6.10. All mismatch-sensing transistors have a shared tail current source I_C. Without mismatches, the currents in all four sensing branches are identical. The voltages V_1–V_4 are also equal in the absence of mismatches since they are derived from comparisons of the drain currents of M_{1S}–M_{4S} with the same current I_P from well-matched current sources with large transistor dimensions. Notice that the current I_P is controlled by a common-mode feedback ($CMFB_{cal}$) loop that regulates the high-impedance nodes at the drains of the sensing-transistors to maintain the average of V_1–V_4 equal to V_{cal}. As in Sect. 6.2, the capacitors C_{st}

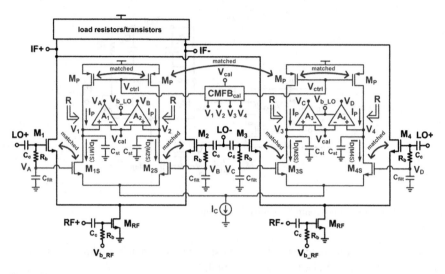

Fig. 6.10 Mixer with calibration loop components

Fig. 6.11 DC signal flow
diagram for one calibration
loop with offsets

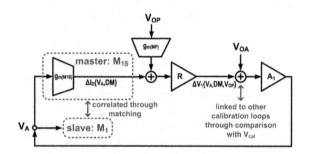

and C_{filt} serve to stabilize the loop and to filter out high-frequency signal
components that might leak into the calibration circuitry. At steady state, the
errors between the currents $I_{D(M1S)}$–$I_{D(M4S)}$ become very small due to the high
loop gain.

With mismatched transistors M_1–M_4 in Fig. 6.10, the different correlated
currents $I_{D(M1S)}$–$I_{D(M4S)}$ of the sensing-transistors will be converted to distinct
voltages V_1–V_4. These voltages are compared to the common-mode voltage V_{cal}
by amplifiers A_1–A_4 in each branch for further amplification and automatic
adjustment of the individual bias voltages V_A–V_D around the set bias V_{b_LO} for the
switching transistors. For example, if $I_{D(M1S)}$ is relatively low compared to the
other currents due to parameter mismatches, then V_1 will be higher than V_{cal}.
Consequently, the output voltage V_A of amplifier A_1 will rise above V_{b_LO}, and the
increase of the gate bias voltage in this branch will increase $I_{D(M1S)}$ until it is equal
to the currents in the other branches.

An equivalent diagram for the DC calibration loop containing M_{1S} is portrayed
in Fig. 6.11, which includes the offsets that affect the scheme's accuracy. It can be

considered a master/slave configuration, in which M_{1S} is in the master loop and the shared gate bias voltage V_A is controlling the slave element M_1. The transconductors $g_{m(M1S)}$ and $g_{m(MP)}$ are representing the transconductance parameters of M_{1S} and M_P in Fig. 6.10. V_{OP} is the gate-referred offset voltage of M_P. The current $\Delta I_D\{V_A, DM\}$ is the difference of the sensing transistor's drain-source current relative to the mean of the same current in all branches, which depends on V_A and the device mismatches (DM) under correction. The block labeled "R" in Fig. 6.11 represents the equivalent resistance looking into the node at which the drains of M_{1S} and M_P are connected together. At this node, the voltage ΔV_1 (the divergence of V_1 from the mean of V_1–V_4) is a function of V_A, V_{OP}, and DM. Furthermore, the input-referred offset voltage V_{OA} of the amplifier A_1 adds at the same node. This node is significant because it links the calibration loop for M_1 to the other branches by comparison of V_1–V_4 with V_{cal} at the inputs of the amplifiers (Fig. 6.10).

As explained in Sect. 6.2, the intrinsic limit of the calibration loop's ability to reduce the standard deviation of the parameter mismatches between the slave transistors in the main circuit depends on their layout-dependent correlation to the mismatch-sensing transistors. For optimum effectiveness, the offsets associated with devices in the loop relative to their counterparts in the other branches must be minimized as well. From Fig. 6.11, two conditions can be identified by inspection:

$$V_{OP} \ll \frac{\Delta I_D\{V_A, DM\}}{g_{m(MP)}} \tag{6.6}$$

$$V_{OA} \ll \Delta I_D\{V_A, DM\} \times R \tag{6.7}$$

Since the offset voltages are inversely proportional to the device dimensions [7] of the current sources M_P and in amplifier A_1, the strategy to meet the criteria in Eqs. 6.6 and 6.7 is to increase these dimensions until the simulated offsets are negligible. This is feasible because the parasitic capacitances from the large devices are not critical in this DC loop.

It is also insightful to assess the input-referred offset voltage of the calibration loop. Since V_A links the master and slave elements, it is preferred to maximize the sensitivity to $\Delta I_D\{V_A, DM\}$ by minimizing the impact of offsets at that node. Referring to Fig. 6.11 again, it can be derived that the offset of V_A (from V_{b_LO} in Fig. 6.10) is:

$$V_A = \frac{\Delta I_D\{V_A, DM\}}{g_{m(M1S)}} + \frac{g_{m(MP)} \cdot V_{OP} + V_{OA}/R}{g_{m(M1S)}} \tag{6.8}$$

Apart from the need to minimize offset voltages V_{OP} and V_{OA}, the above expression reveals the importance of maximizing the gain in the first amplification stage by designing R to be large. This suggests the use of a small current I_C in combination with non-minimum transistor lengths for M_P to increase the resistance looking into the node at the drains of M_P and M_{1S} in Fig. 6.10.

Since the nodes labeled V_1–V_4 in Fig. 6.10 are high-impedance nodes, common-mode control circuitry is necessary to ensure that the positive inputs of

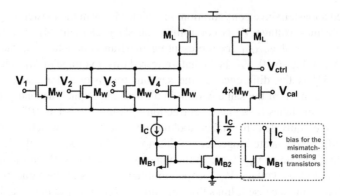

Fig. 6.12 Common-mode feedback circuit for the main calibration loop

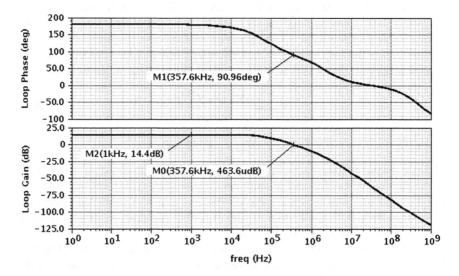

Fig. 6.13 Frequency response of the main CMFB circuit

amplifiers A_1–A_4 are maintained close to the calibration reference V_{cal} at the negative inputs. Figure 6.12 shows the schematic of the CMFB circuit in the calibration loop, which weighs voltages V_1–V_4 equally and compares their averaged value to the reference voltage V_{cal}. For convenience, the current mirror to bias the CMFB circuit also provides the current I_C that is routed to the sources of the mismatch-sensing transistors in the main loop. The stability of the CMFB loop is strongly related to the main calibration loop due to the shared dominant pole at V_1–V_4. Hence, a large value can be selected for C_{st} in Fig. 6.10 in order to stabilize both loops. A mixer calibration design example will be described with more details in the remainder of this chapter. The simulated gain and phase

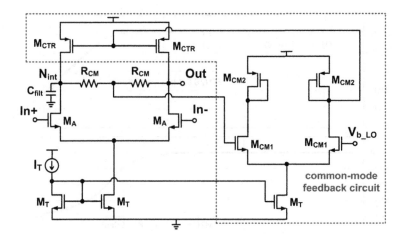

Fig. 6.14 Schematic of amplifiers A_1–A_4 in the calibration loop

responses of the CMFB loop in this design are displayed in Fig. 6.13. It has a low-frequency gain of 14.4 dB and a phase margin of 91.0°.

The schematic of the amplifiers A_1–A_4 is displayed in Fig. 6.14. It consists of a simple differential pair (M_A) loaded by resistors (R_{CM}) and controlled current sources (M_{CTR}). The resistors serve as common-mode detectors for the CMFB amplifier (M_{CM1}, M_{CM2}) that is connected to the gates of M_{CTR} to regulate the output of the main amplifier. When the mismatches are sensed (In+ to In− $\neq$ 0), the voltage at the output terminal (Out) can move freely to counteract the sensed difference as part of the mismatch calibration loop, but the CMFB of the amplifier ensures that this change occurs around the required gate bias voltage level V_{b_LO} of the switching transistors in the mixer. Besides its role in the common-mode detection, the internal node N_{int} is not utilized as output. However, the same capacitor C_{filt} as present at the amplifier output (Fig. 6.10) is connected to N_{int} for loading symmetry.

In this example, the amplifier in Fig. 6.14 was designed with a DC gain of 21.5 dB from the differential input to the single-ended output. Apart from the stability considerations, its frequency response (Fig. 6.15) is not critical in the DC calibration loop for static mismatch reduction. Nonetheless, the bandwidth of the amplifier and overall calibration loop can be optimized when fast settling is desired for test time reduction.

To demonstrate the calibration method, a double-balanced mixer was designed (see Sect. 6.3.3 for details) with the auxiliary circuitry described above. Table 6.2 lists the component parameters of the design in TSMC 0.13 μm CMOS technology using a 1.2 V supply. Only the mismatch-sensing transistors M_{1S}–M_{4S} have minimum transistor lengths. Their dimensions were selected identical to those of the switching transistors in the mixer under calibration, and they have the same number of fingers for improved parameter correlations according to Eqs. 6.1 and 6.2. As explained previously, all other transistors in the mismatch calibration loop have non-minimum dimensions to decrease mismatches and offset voltages.

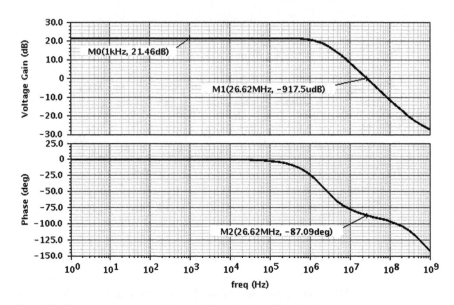

Fig. 6.15 Frequency response of the amplifiers in the calibration loop

Table 6.2 Calibration circuitry components (0.13 μm CMOS technology with 1.2 V supply)

Component	Dimensions/Value
Mismatch-sensing and first amplification stage (Fig. 6.10):	
M_{1S}, M_{2S}, M_{3S}, M_{4S}	W/L = 2 μm × 40 fingers/0.13 μm
M_P	W/L = 6.9 μm × 12 fingers/5 μm
C_{st}	55 pF
C_{filt}	0.5 pF
V_{cal}	0.8 V
I_C	50 μA
V_{b_LO}	0.665 V
C_c	1 pF
R_b	100 kΩ (L/W = 6 × 15.8/1 μm)
Common-mode feedback circuit (Fig. 6.12):	
M_W	W/L = 3 μm × 4 fingers/0.3 μm
M_L	W/L = 3.3 μm × 2 fingers/0.3 μm
M_{B1}	W/L = 2.5 μm × 8 fingers/0.5 μm
M_{B2}	W/L = 2.5 μm × 4 fingers/0.5 μm
Amplifiers A_1–A_4 (Fig. 6.14):	
M_A	W/L = 6 μm × 14 fingers/4 μm
M_{CTR}	W/L = 5.2 μm × 8 fingers/3 μm
M_{CM1}	W/L = 1.8 μm × 2 fingers/0.3 μm
M_{CM2}	W/L = 2.8 μm × 2 fingers/0.3 μm
M_T	W/L = 2 μm × 8 fingers/1 μm
I_T	20 μA
R_{CM}	128 kΩ (L/W = 20 × 6/1 μm)

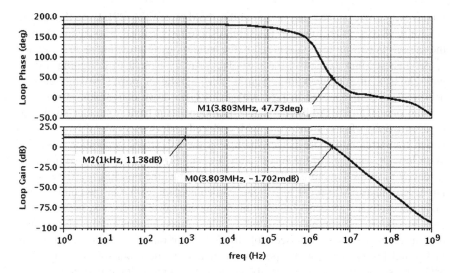

Fig. 6.16 Open-loop frequency response of the calibration circuit. (The simulation for a single branch was performed with the CMFB$_{cal}$ block activated.)

With the design parameters in Table 6.2, the DC gain from the gate to the drain of each sensing transistor (M$_{1S}$–M$_{4S}$) is 20.5 dB. Considering the 21.5 dB amplifier gain (A$_1$–A$_4$), the total DC loop gain per branch is 42 dB. When assessing the stability, it is important to keep in mind that the loops interact through the shared sources of M$_{1S}$–M$_{4S}$ and the common-mode feedback circuit (CMFB$_{cal}$). Simulations were performed to determine the appropriate capacitor values of C$_{st}$ and C$_{filt}$ for stability by inserting a probe at the gate of one mismatch-sensing transistor in Fig. 6.10 and plotting the loop's frequency response, which is also influenced by the CMFB$_{cal}$ circuit. This assessment is to assure tolerance to any perturbation that could occur from high-frequency noise in one branch. The gain of the differential comparison involving the mismatch currents in each branch is very high, which is also evident from the evaluation of the mismatch current reduction that follows in Sect. 6.3.4. Nevertheless, the response to an AC disturbance in an individual loop has a lower gain when only one of the voltage inputs (V$_1$–V$_4$) of the CMFB$_{cal}$ block changes because the common-mode feedback action lowers the single-ended equivalent impedance seen at nodes V$_1$–V$_4$. As shown in Fig. 6.16, this combined loop response for a single branch has an effective DC gain of 11.4 dB and phase margin of 47.7° at the 3.8 MHz unity gain frequency.

6.3.3 Double-Balanced Mixer Design

Since the transition frequency (f$_T$) of devices in CMOS technologies continues to increase, several recent works have taken advantage of this trend by designing RF mixers with devices operating in the subthreshold region [26–29]. Even though the f_T of a device is much lower in the subthreshold (weak inversion) region than in the saturation (strong inversion) region, the technology improvements make up for f_T deficiencies that existed in the past. The primary benefit of designing mixers with devices in subthreshold region is that significant power savings can be achieved, as demonstrated in [26] with a 2.4 GHz down-conversion mixer consuming only 0.5mW. Additionally, the LO signal can have a smaller swing for hard-switching of the transistors with reduced gate-source overdrive voltage, which translates into more power savings in the LO signal generation circuitry. With less DC currents in the mixer branches, subthreshold designs also have the tendency to allow for more voltage headroom. Thus, the possibility exists to use larger load resistors in order to increase the conversion gain. On the contrary, the main trade-offs are reduced linearity, higher device noise levels, and increased die area to obtain comparable transconductance values. Furthermore, subthreshold designs are generally more susceptible to PVT variations. For example, the results in [2] and [30] show how the percent mismatch of the drain-source current for MOS transistors increases drastically as the gate-source voltage is decreased.

Although the IIP2 calibration technique presented in the previous subsection can be applied to any double-balanced mixer, it is demonstrated here for a subthreshold mixer in order to simultaneously explore this promising design methodology further. Figure 6.17 shows the mixer schematic from before with more details. The approach taken here is to optimize the subthreshold mixer for linearity and noise performance that approximates state of the art mixers in saturation region as much as possible for typical conversion gain. This requires transistors with high W/L ratios to obtain the appropriate transconductances in subthreshold region. However, the use of large devices increases the total parasitic capacitances at the drains of the LO transistors (M_{SW}), causing IIP2 degradation. As explained in [11], the inductors (L_S) resonate with these parasitic capacitances to improve the IIP2 performance. In addition, the mismatch reduction method for the LO transistors is utilized for further IIP2 enhancement. While the LO transistor bias voltages V_A–V_D are generated with the previously described loop, the RF input transconductors are biased with a simple current mirror to produce the DC current I_{DC} on each side of the mixer. If the transconductance mismatch of the M_{RF} transistors becomes detrimental, then the same mismatch reduction loop as for the LO transistors can be employed to generate the RF bias voltages individually. However, IIP2 is typically more sensitive to LO transistor mismatches as described in Sect. 6.3.1. To achieve sufficient transconductance in this subthreshold mixer design, the RF input transistors M_{RF} are five times larger than the LO transistors M_1–M_4, which makes it even less important to calibrate the M_{RF} transistors.

As the subthreshold mixers in [28, 29], the mixer in Fig. 6.17 has an active load consisting of transistors (M_{ctrL}) and resistors (R_L). The capacitor C_L represents the input capacitance of the following filter or output buffer stage. A common-mode feedback loop (CMFB) with relatively high gain over the IF signal bandwidth is

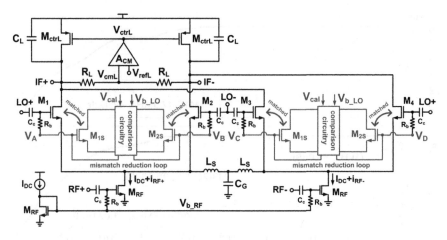

Fig. 6.17 Detailed double-balanced mixer schematic

Fig. 6.18 Common-mode feedback amplifier at the mixer output

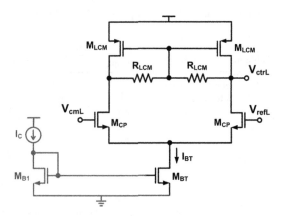

employed at the mixer output, which regulates the DC output voltage level around V_{refL} and aids by suppressing the common-mode IM2 components [20]. The amplifier A_{CM} in this CMFB loop is displayed in Fig. 6.18. This amplifier is a simple differential pair with self-regulated active load. Its bias current provided by transistor M_{BT} is obtained from the gate voltage of the diode-connected transistor in the core calibration circuitry (M_{B1} in Fig. 6.12). The simulated frequency response of the output CMFB loop is shown in Fig. 6.19, revealing high low-frequency gain of 35 dB as well as 26 dB at 20 MHz to cover a wide IF signal bandwidth.

Table 6.3 lists the component dimensions and values of key design parameters for the mixer and its auxiliary circuitry. Notice that the dimensions and number of fingers of the switching transistors M_1–M_4 are exactly the same as the sensing transistors M_{1S}–M_{4S} in the mismatch reduction loop.

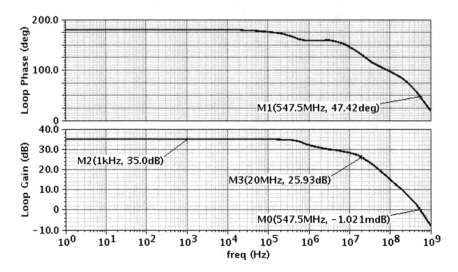

Fig. 6.19 Simulated gain and phase of the CMFB loop at the mixer output

Table 6.3 Subthreshold mixer components (0.13 μm CMOS technology with 1.2 V supply)

Component	Dimensions/Value
Main mixer components (Fig. 6.17):	
M_1, M_2, M_3, M_4	W/L = 2 μm × 40 fingers/0.13 μm
M_{RF}	W/L = 10 μm × 40 fingers/0.13 μm
M_{ctrL}	W/L = 1.2 μm × 26 fingers/0.25 μm
R_L	3 kΩ (L/W = 10 × 8.87/8 μm)
C_L	0.15 pF
L_S	7 nH
C_c	1 pF
R_b	100 kΩ (L/W = 6 × 15.8/1 μm)
V_{b_LO} (nominal values of V_A, V_B, V_C, V_D)	0.665 V
V_{refL}	0.565 V
I_{DC}	200 μA
Common-mode feedback amplifier A_{CM} (Fig. 6.18):	
M_{CP}	W/L = 1.5 μm × 4 fingers/0.13 μm
M_{LCM}	W/L = 1.5 μm × 4 fingers/0.13 μm
M_{B1}	W/L = 2.5 μm × 8 fingers/0.5 μm
M_{BT}	W/L = 2.5 μm × 18 fingers/0.5 μm
R_{LCM}	3.9 kΩ (L/W = 6 × 4.5/2 μm)
I_{BT}/I_C	110/50 μA
Mismatch reduction loop (Fig. 6.10):	
M_{1S}, M_{2S}, M_{3S}, M_{4S}, comparison circuitry	listed in Table 6.2

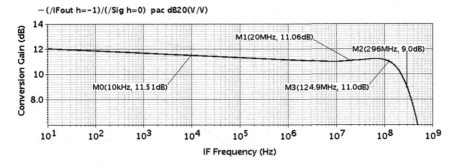

Fig. 6.20 Conversion gain vs. frequency

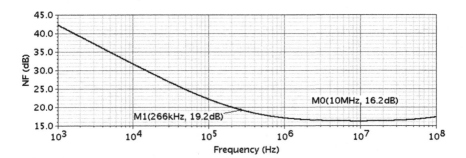

Fig. 6.21 SSB noise figure vs. frequency

6.3.4 Simulation Results

6.3.4.1 Characterization of the Subthreshold Mixer Design

Unless noted otherwise, the simulation results for the subthreshold mixer design described in Sect. 6.3.3 were obtained with a 1.988 GHz sinusoidal LO signal having a power of −1 dBm. As seen in Fig. 6.20, this mixer has a conversion gain of 11.5 dB ± 0.5 dB for RF input signals located up to 125 MHz away from the LO frequency. It has been demonstrated that designing active mixers in the subthreshold region allows high gain (e.g., 32 dB in [29]) with low power consumption from the use of small bias currents, which also leaves voltage headroom for large load resistors. However, the mixer in this example was optimized to achieve high linearity for broadband applications. This required a conversion gain trade-off that resulted in 11.5 dB gain, which is comparable to conventional double-balanced active mixers designed with transistors in the saturation region.

Figure 6.21 shows that a reasonable noise figure (NF) can be attained in the subthreshold region by using large RF input transistors to ensure that they have sufficient transconductance. In this case, the single-sideband (SSB) NF is 16.2 dB

Fig. 6.22 IIP3 curve.
LO frequency: 1.988 GHz,
RF test tones: 2 GHz,
2.004 GHz, IM3 frequency:
8 MHz

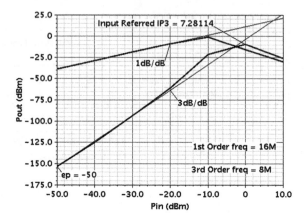

Fig. 6.23 1 dB compression
curve

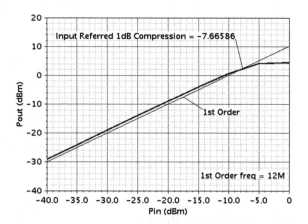

with a flicker noise corner at 266 kHz. The corresponding double-sideband (DSB) NF is normally 3 dB lower than the SSB NF [15].

Linearity characteristics were assessed within a 20 MHz band under consideration that the mixer is intended for broadband wireless target application such as WiMAX. The simulated IIP3 of 7.3 dBm in Fig. 6.22 was obtained with two tones located at 2 GHz and 2.004 GHz (12 MHz and 16 MHz away from the 1.988 GHz LO frequency). Figure 6.23 shows that the mixer has a simulated 1 dB compression point of −7.7 dBm, which was determined by sweeping the power of a single 2 GHz RF input tone.

To give first insights into the IIP2 characteristics, the simulated IIP2 curves with 0.5% load resistor mismatches are plotted in Fig. 6.24 for the mixer without and with calibration circuitry. This assessment condition was selected because the load mismatch leads to common-mode to differential-mode conversion of the IM2 components according to Eq. 6.4. Without any other mismatches in the circuits,

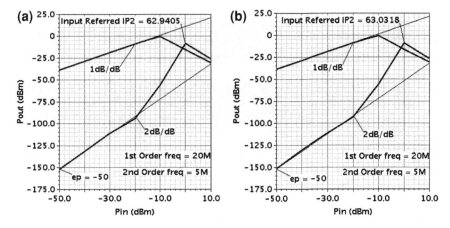

Fig. 6.24 IIP2 curve with 0.5% mismatch between the load resistors (R_L). LO frequency: 1.985 GHz, RF test tones: 2 GHz, 2.005 GHz, IM2 frequency: 5 MHz; **a** without calibration circuitry, **b** with calibration circuitry

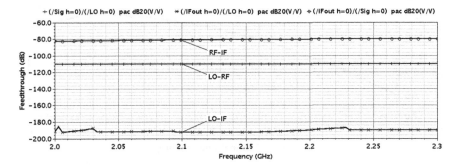

Fig. 6.25 Feedthrough between mixer ports

the results in Fig. 6.24 reveal that the calibration circuitry has negligible impact. IIP2 characterizations with Monte Carlo simulations using statistical device models provided by the foundry are discussed later in this section to present an estimate for the IIP2 improvement from the calibration circuit in the presence of realistic device mismatches in the mixer and calibration circuit itself.

Figure 6.25 displays the simulated port–port feedthroughs, showing that the port–port isolation is 80 dB or more. This isolation is credited to the fact that minimum lengths are used for the LO switching transistors and RF input transistors, which is particularly important to minimize the parasitic capacitances when designing in the subthreshold region with high W/L ratios. As for conventional mixers, the measured isolation will be strongly affected by substrate leakage and layout parasitics, as well as package and PCB design choices. As explained in Sect. 6.2, one of the motivations behind the use of the DC calibration loop with

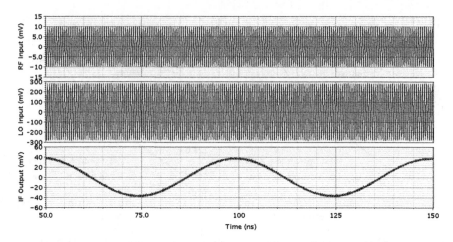

Fig. 6.26 Transient simulation with a 20 MHz IF output signal. (LO frequency: 1.985 GHz, RF input signal: −30 dBm at 2.005 GHz.)

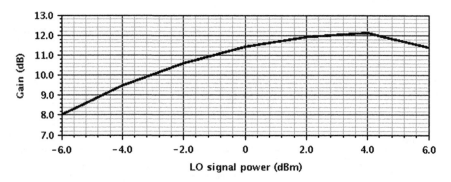

Fig. 6.27 Conversion gain vs. LO signal power. (frequencies: LO = 1.985 GHz, RF 2.005 GHz, IF = 20 MHz.)

low-pass filter nodes is to avoid RF coupling and substrate leakage due to the proximity of transistors in typical layout matching techniques.

Figure 6.26 shows the transient signals from a simulation of the mixer with a −30 dBm differential RF input signal at 2.005 GHz and a −1 dBm differential LO at 1.985 GHz. As expected, the differential IF output signal (IF+ to IF−) has a frequency of 20 MHz and an amplitude of 38.8 mV, indicating a conversion gain of 11.8 dB relative to the 10 mV RF input amplitude.

Since the mixer is designed in the subthreshold region instead of the saturation region, a smaller LO amplitude is needed to induce hard-switching of the LO transistors due to the reduced gate-source overdrive voltage. The progression of the simulated gain, NF, IIP2, and IIP3 for a sweep of the LO signal power can be observed in Figs. 6.27, 6.28, 6.29. Based on the specification trade-offs in these plots, the LO power of −1 dBm was selected for this subthreshold mixer design.

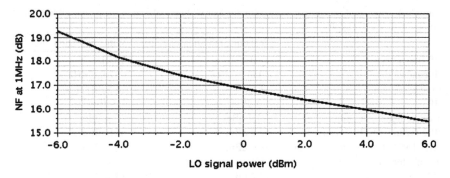

Fig. 6.28 SSB noise figure at IF = 1 MHz vs. LO signal power

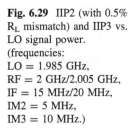

Fig. 6.29 IIP2 (with 0.5%
R_L mismatch) and IIP3 vs.
LO signal power.
(frequencies:
LO = 1.985 GHz,
RF = 2 GHz/2.005 GHz,
IF = 15 MHz/20 MHz,
IM2 = 5 MHz,
IM3 = 10 MHz.)

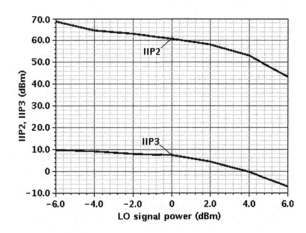

A summary of the subthreshold mixer performance specifications is provided in
Table 6.4 to compare the simulation results before and after adding the calibration
circuitry. The outcomes show that none of the mixer specifications is affected
significantly by the DC calibration loops outside of the signal path. A notable
difference is the minimum IIP2 observed after Monte Carlo simulations, which
will be discussed in the remainder of this chapter. In general, the impact of the
mixer's auxiliary calibration circuits is limited to its ability to compensate for
device variations and mismatches as discussed in Sects. 6.1 and 6.2. However, the
drawbacks are the increase of the total power consumption from 0.68 to 0.97 mW
as well as the die area required for the calibration circuitry.

6.3.4.2 IIP2 Evaluation Before and After the Addition
of the Calibration Circuitry

The IIP2 performance was investigated with statistical Monte Carlo simulations
using device models provided by the foundry to account for process and mismatch

Table 6.4 Simulated mixer specifications with and without calibration (0.13 μm CMOS technology with 1.2 V supply)

	Without calibration circuitry	With calibration circuitry
RF frequency	2 GHz	2 GHz
IF bandwidth	<124.9 MHz	<124.3 MHz
Conversion gain	11.5 dB	11.5 dB
IIP3	7.3 dBm	7.3 dBm
1-dB compression point	−7.7 dBm	−7.8 dBm
IIP2 (With 0.5% R_L mismatch)	62.9 dBm	63.0 dBm
Avg. IIP2[a] (100 Monte carlo runs)	58.9 dBm	64.2 dBm
Yield[b] (for IIP2 > 54dBm)	75%	91%
DSB noise figure	13.2 dB	13.2dB
Flicker noise corner	266 kHz	274 kHz
LO-RF isolation (2–2.3 GHz)	>110 dB	>110 dB
LO-IF isolation (2–2.3 GHz)	>185 dB	>182 dB
RF-IF isolation (2–2.3 GHz)	>80 dB	>79 dB
Power (with auxiliary circuits)	0.68 mW	0.97 mW

[a] With foundry-supplied statistical models (process and mismatch) for all devices in the mixer and calibration circuits
[b] Defined as the percentage of the Monte Carlo simulation outcomes that meet the IIP2 target

variability. All active and passive devices in the mixer and calibration circuit were simulated with these statistical models, and correlations between matched devices were defined based on Eqs. 6.1 and 6.2 as described in Sects. 6.2.1 and 6.2.2. In the mixer, correlations based on the number of fingers or resistor segments were set only for the load devices R_L and M_{ctrL} in Fig. 6.17 as well as the devices with identical names in the CMFB circuit in Fig. 6.18. This was done under the assumptions that these will be laid out with matching techniques. On the contrary, correlations were not specified for the devices that process RF signals (M_1–M_4 and M_{RF}), so that these can be placed as individual devices to minimize substrate leakage due to placement proximity and crosstalk via routing parasitics. Since parasitic capacitances in the low-frequency calibration circuits are not critical, they can be laid out with matching techniques. Hence, correlations were defined based on the number of fingers or resistor segments for M_{1S}–M_{4S} and M_P in Fig. 6.10 as well as for the transistors and resistors with equal labels in the CMFB$_{cal}$ (Fig. 6.12) and amplifier circuits (Fig. 6.14).

Figure 6.30 displays the histograms of the IIP2 from Monte Carlo simulations (process and mismatch variations enabled) with 100 runs before and after the addition of the calibration circuitry. Without calibration, the IIP2 mean is 58.9 dBm (with 7.6 dBm standard deviation), which improved to 64.2 dBm (with 8.7 dBm standard deviation) due to the calibration. With a target IIP2 of 54 dBm for example, this would correspond to a yield increase from 75 to 91% as a result of the calibration.

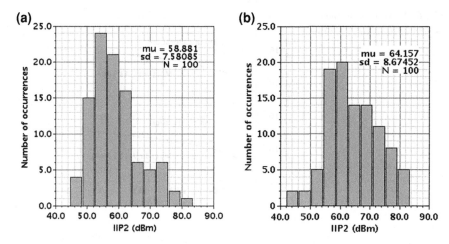

Fig. 6.30 IIP2 comparison with 100 Monte Carlo runs. LO frequency: 1.985 GHz, RF test tones: 2 GHz, 2.005 GHz, IM2 frequency: 5 MHz; **a** without calibration circuitry, **b** with calibration circuitry

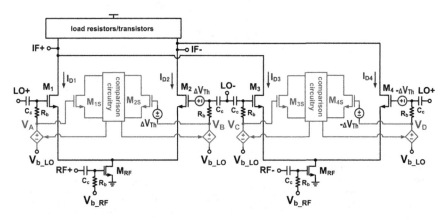

Fig. 6.31 Mixer with intentional threshold voltage offsets (ΔV_{Th})

6.3.4.3 Mismatch Reduction with the Calibration Loops

The mismatch in the mixer core can be assessed by purposely introducing offset voltages at the gates of the LO transistors to emulate threshold voltage mismatches as visualized in Fig. 6.31. In this test setup, a positive DC offset voltage source (ΔV_{Th}) was inserted at the gates of M_2 and to its corresponding matched sensing-transistor M_{2S}, while the same offset voltage with negative polarity was included at the gates of M_4 and M_{4S}. The ultimate mismatch indicator is the difference of the LO transistor DC drain currents I_{D1}–I_{D4}. Here, this average mismatch current is defined using I_{D1} as reference:

Fig. 6.32 Average mismatch current (of I_{D1}–I_{D4}) vs. ΔV_{Th}

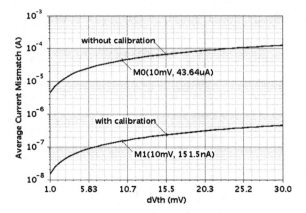

Fig. 6.33 Transient settling behavior of critical control voltages

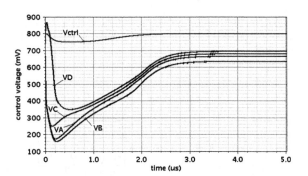

Fig. 6.34 Transient IF output after settling of the calibration control voltages (LO frequency: 1.985 GHz, RF input signal: −30 dBm at 2.005 GHz.)

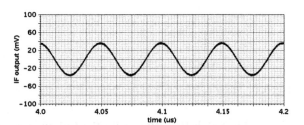

$$\overline{\Delta I_D} = mean\{|I_{D2} - I_{D1}| + |I_{D3} - I_{D1}| + |I_{D4} - I_{D1}|\} \qquad (6.9)$$

A comparison of the average mismatch current with and without calibration circuitry is plotted in Fig. 6.32 for a sweep of ΔV_{Th} from 1 to 30 mV, showing a mismatch current reduction by more than two orders of magnitude. This property of the calibrated mixer is the fundamental mechanism behind the IIP2 improvement observed in the Monte Carlo simulation results.

Fig. 6.35 Conversion gain comparison with 100 Monte Carlo runs. LO frequency: 1.98 GHz, RF test tone: 2 GHz, IF frequency: 20 MHz; **a** without calibration circuitry, **b** with calibration circuitry

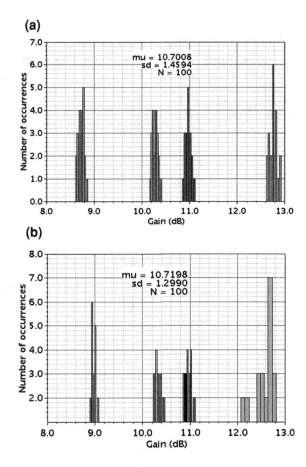

6.3.4.4 Transient Behavior of the Calibration Loops

Figure 6.33 shows the settling of voltages V_A–V_D and V_{ctrl} (Fig. 6.10) from a transient simulation with 1.985 GHz LO frequency and a -30 dBm RF input signal at 2.005 GHz. In this simulation, the offset voltages at the gates of (M_2, M_3, M_4) changed from 0 V to (30, -15, -30 mV) at time $= 0$ s. Figure 6.34 displays the corresponding transient waveform of the down-converted 20 MHz signal at the IF output after settling of the control voltages. The short settling times below 4 μs of the control voltages in this background calibration scheme make it suitable for quick calibrations at system start-up as well as for in built-in test routines during manufacturing testing.

Fig. 6.36 IIP3 comparison
with 100 Monte Carlo runs.
LO frequency: 1.98 GHz, RF
test tones: 2.01 GHz,
2.02 GHz, IM3 frequency:
20 MHz; **a** without
calibration circuitry,
b with calibration circuitry

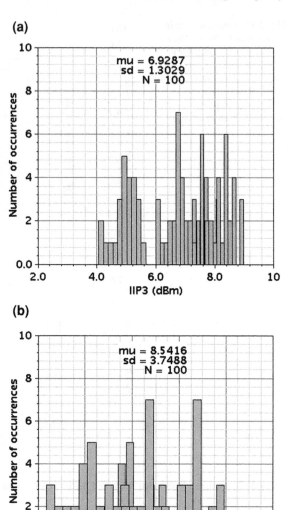

6.3.4.5 Variations of Other Mixer Parameters

Monte Carlo simulations with statistical device models and the aforementioned
correlation definitions were also performed to determine the calibration circuitry's
impact on other key mixer specifications. Figures 6.35, 6.36, 6.37 show the histograms of the conversion gain, IIP3, and noise figure after Monte Carlo simulations with 100 runs. By comparing the results, it can be seen that the calibration
circuitry has little impact on the mean values and standard deviations of these

Fig. 6.37 Comparison of the SSB NF at 1 MHz with 100 Monte Carlo runs. The shown cases are: **a** without calibration circuitry, **b** with calibration circuitry

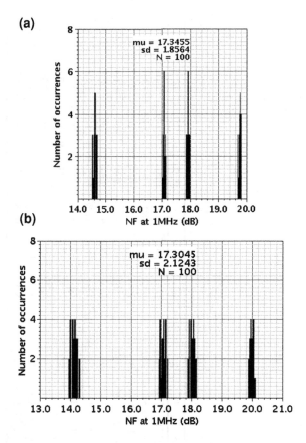

specifications. However, activation of the calibration slightly increases the IIP3 and its standard deviations by 1.6 and 2.5 dBm, respectively.

6.3.4.6 Assessment with Respect to the State of the Art

Table 6.5 contains summaries of specifications reported for CMOS down-conversion mixers with similar operating frequencies. The presented subthreshold mixer in the last column has lower IIP2 than the mixers in columns 1–6 that are designed with transistors biased in saturation region. However, when the IIP2 target is 50 dBm as in [13], the IIP2 improvement from the calibration makes it possible to achieve such a target with this subthreshold design. Apart from mixer design optimizations for scenarios with higher IIP2 requirement, it can be explored to make the load resistors of the mixer programmable for further IIP2 tuning through digital trimming as proposed in [16]. Most of the mixers in columns 1–6 of Table 6.5 contain auxiliary circuitry for IIP2 enhancements. Notice that they exhibit overall comparable performances but consume at least six times as much power as the discussed subthreshold mixer with calibration. On the other hand, the

Table 6.5 Down-conversion mixer performance comparison

	1	2	3	4	5	6	7	8	9	10	11
Reference	[3]$^{\Delta *}$	[11]$^{\Delta\dagger}$	[13]†	[14]$^{\Delta\dagger}$	[23]$^{\Delta\dagger}$	[24]$^{\Delta *}$	[26]$^{\#\dagger}$	[27]$^{\#\dagger}$	[28]$^{\#\#}$	[29]$^{\#\dagger}$	This example$^{\#\Delta *}$
CMOS technology (nm)	180	180	90	350	130	65	130	180	130	180	130
RF Freq. (GHz)	3.5	2.1	2.1	0.815	2	2.1	2.4	2.4	3.1–10.6	2.4	2
IF freq. (MHz)	–	<4.5	<1.2	<10	<1.5	<10	60	10	264	30	<124
Conversion gain (dB)	10	16	9	14.5	53	8	15.7	9	9.8–14.0	32	11.5
Noise Meas. or DSB NF (dB)	4.5 nV/$\sqrt{\text{Hz}}^{\S}$	4 nV/$\sqrt{\text{Hz}}^{\S}$	9.4	12	3.5 nV/$\sqrt{\text{Hz}}^{\S}$	16	18.3	11.8	14.5–19.6	8.5	13.2
IIP3 (dBm)	8	9	8.9	2.4	12	12	–9	–	–11	–14.5	7.3
1-dB Comp. Point (dBm)	–	–	–	–	4	–	–28	–	–24 – –19	–	–7.8
IIP2 (dBm)	>65	>78	>55.1	>66	~85	>75	–	–	–	–	>54^{X}
Supply (V)	1.8	1.8	1	2.7	1.5	1	1	1.2	1.2	1.8	1.2
Power (mW)	–	7.2	6.25$^{\theta}$	10.8	72	8.5	0.5	0.18	1.85	1	0.97

* Simulation results; † Measurement results; # Measurement results; $^{\Delta}$ With IIP2 enhancement circuitry; $^{\theta}$ Reported with LO buffer; § Reported as input-referred noise; ^{X} With 91% yield

subthreshold mixer designs in columns 7–10 have similar performances and power consumptions compared to this example design, but with the tendency that they have lower IIP3 and 1 dB compression point specifications; whereas IIP2 characterization results were not reported for these designs. In general, the presented subthreshold mixer with calibration has competitive performance relative to saturation region mixers, but with significantly lower power dissipation in the same range as other reported subthreshold mixers. The simulation results suggest that the calibration loop effectively improves the second-order linearity and makes the subthreshold design more robust to mismatch variations.

6.4   Summarizing Remarks

Alternatively to matching transistors within the RF signal path or increasing their dimensions, a methodology has been described to reduce the mismatch between a pair of transistors by indirectly matching them through a DC calibration loop. Monte Carlo simulation results demonstrated that the input offset standard deviation of the differential amplifier under investigation is expected to reduce from 4.17 to 1.29 or 0.76 mV, which depends on the layout-based quality of the matching between the RF and mismatch-sensing transistors. The trade-offs with the scheme are an approximately 15% power increase and the die area overhead for the calibration circuitry.

Applied to an example mixer design, it was shown that the calibration scheme improves the IIP2 specification. Monte Carlo simulations revealed that the mean of the IIP2 increased from 58.9 to 64.2 dBm. While the background calibration loops did not noticeably impact other mixer specifications, the main trade-off was a 30% increase in the power consumption. If the mixer under calibration is designed with saturation region bias conditions using higher currents, then the power overhead could be as low as 10–20% because the bias currents in the amplifiers of the calibration loop can be maintained small. The other investment with this IIP2 enhancement method is the die area required for the calibration circuitry. Depending on the layout style, the mixer area with calibration could be up to twice the area of the mixer without calibration. There is a direct trade-off between the layout area and the IIP2 improvement from better matching between devices. But unlike with conventional matching techniques, the devices with non-minimum lengths in the calibration loops are outside of the signal path and therefore their parasitic capacitances do not degrade the mixer's frequency response.

References

1. D. Gomez, D. Mateo, Exploiting CMOS short-channel effects for yield enhancement in analogue/RF design. Electron. Lett. **46**(8), 559–561 (2010)
2. K. Agarwal, J. Hayes, S. Nassif, Fast characterization of threshold voltage fluctuation in MOS devices. IEEE Trans. Semicond. Manuf. **21**(4), 526–533 (2008)

3. S. Rodriguez, A. Rusu, L.-R. Zheng, M. Ismail, CMOS RF mixer with digitally enhanced IIP2. Electron. Lett. **44**(2), 121–122 (2008)
4. V. Gupta, G.A. Rincon-Mora, Achieving less than 2% 3-σ mismatch with minimum channel-length CMOS devices. IEEE Trans. Circuits Syst. II: Express Briefs **54**(3), 232–236 (2007)
5. M. Conti, P. Crippa, S. Orcioni, C. Turchetti, Layout-based statistical modeling for the prediction of the matching properties of MOS transistors. IEEE Trans. Circuits Syst. I: Fundam. Theor. App. **49**(5), 680–685 (2002)
6. T.-H. Yeh, J.C.H. Lin, S.-C. Wong, H. Huang, J.Y.C. Sun, Mis-match characterization of 1.8 and 3.3 V devices in 0.18 μm mixed signal CMOS technology. in *Proceedings IEEE International Conference Microelectronic Test Structures (ICMTS)*, March 2001, pp. 77–82
7. M.J.M. Pelgrom, A.C.J. Duinmaijer, A.P.G. Welbers, Matching properties of MOS transistors. IEEE J. Solid-State Circuits **24**(5), 1433–1439 (1989)
8. Cadence Design Systems, Recommended Monte Carlo modeling methodology for Virtuoso Spectre circuit simulator application note, Nov. 2003, pp. 13–18. Available: http://www.cdnusers.org/community/virtuoso/resources/spectre_mcmodelingAN.pdf
9. K. Kivekas, A. Parssinen, K.A.I. Halonen, Characterization of IIP2 and DC-offsets in transconductance mixers. IEEE Trans. Circuits Syst. II: Analog Digital Signal Process. **48**(11), 1028–1038 (2001)
10. D. Manstretta, M. Brandolini, F. Svelto, Second-order intermodulation mechanisms in CMOS downconverters. IEEE J. Solid-State Circuits **38**(3), 394–406 (2003)
11. M. Brandolini, P. Rossi, D. Sanzogni, F. Svelto, A +78 dBm IIP2 CMOS direct downconversion mixer for fully integrated UMTS receivers. IEEE J. Solid-State Circuits **41**(3), 552–559 (2006)
12. S. Rodriguez, S. Tao, M. Ismail, A. Rusu, An IIP2 digital calibration technique for passive CMOS down-converters. in *Proceedings of IEEE International Symposium Circuits and Systems (ISCAS)*, May 2010, pp. 825–828
13. S. Peng, C.-C. Chen, and A. Bellaouar, A wide-band mixer for WCDMA/CDMA2000 in 90 nm digital CMOS process. in *Proceedings of IEEE Radio Frequency Integrated Circuits (RFIC) Symposium*, June 2005, pp. 179–182
14. E.E. Bautista, B. Bastani, J. Heck, A high IIP2 downconversion mixer using dynamic matching. IEEE J. Solid-State Circuits **35**(12), 1934–1941 (2000)
15. T.H. Lee, *The Design of CMOS Radio-Frequency Integrated Circuits* (Cambridge University Press, Cambridge, 1998)
16. K. Kivekas, A. Parssinen, J. Ryynanen, J. Jussila, K. Halonen, Calibration techniques of active BiCMOS mixers. IEEE J. Solid-State Circuits **37**(6), 766–769 (2002)
17. W. Kim, S.-G. Yang, J. Yu, H. Shin, W. Choo, B.-H. Park, A direct conversion receiver with an IP2 calibrator for CDMA/PCS/GPS/AMPS applications. IEEE J. Solid-State Circuits **41**(7), 1535–1541 (2006)
18. M. Hotti, J. Ryynanen, K. Halonen, IIP2 calibration methods for current output mixer in direct-conversion receivers. in *Proceedings of IEEE International Symposium Circuits and Systems (ISCAS)*, vol. 5, May 2005, pp. 5059–5062
19. K. Dufrene and R. Weigel, "A novel IP2 calibration method for low-voltage downconversion mixers," in *Proc. IEEE Radio Frequency Integrated Circuits (RFIC) Symposium*, June 2006, pp. 327-330
20. M. Brandolini, M. Sosio, F. Svelto, A 750 mV fully integrated direct conversion receiver front-end for GSM in 90-nm CMOS. IEEE J. Solid-State Circuits **42**(6), 1310–1317 (2007)
21. H. Darabi, H. J. Kim, J. Chiu, B. Ibrahim, L. Serrano, An IP2 improvement technique for zero-IF down-converters. in *IEEE International Solid-State Circuits Conference (ISSCC) Digest of Technical Papers*, Feb 2006, pp. 1860–1869
22. M. Chen, Y. Wu, and M. F. Chang, Active 2nd-order intermodulation calibration for direct-conversion receivers. in *IEEE International Solid-State Circuits Conference (ISSCC) Digest of Technical Papers*, Feb 2006, pp. 1830–1839
23. K. Dufrene, Z. Boos, R. Weigel, Digital adaptive IIP2 calibration scheme for CMOS downconversion mixers. IEEE J. Solid-State Circuits **43**(11), 2434–2445 (2008)

24. M.B. Vahidfar, O. Shoaei, A New IIP2 enhancement technique for CMOS down-converter mixers. IEEE Trans. Circuits Syst. II: Express Briefs **54**(12), 1062–1066 (2007)
25. P. Sivonen, A. Vilander, A. Parssinen, Cancellation of second-order intermodulation distortion and enhancement of IIP2 in common-source and common-emitter RF transconductors. IEEE Trans. Circuits Syst. I: Regul. Pap. **52**(2), 305–317 (2005)
26. H. Lee and S. Mohammadi, A 500 µW 2.4 GHz CMOS subthreshold mixer for ultra low power applications. in *Proceedings of IEEE Radio Frequency Integrated Circuits (RFIC) Symposium*, June 2007, pp. 325–328
27. B.G. Perumana, R. Mukhopadhyay, S. Chakraborty, C.-H. Lee, J. Laskar, A low-power fully monolithic subthreshold CMOS receiver with integrated LO generation for 2.4 GHz wireless PAN applications. IEEE J. Solid-State Circuits **43**(10), 2229–2238 (2008)
28. J.-B. Seo, J.-H. Kim, H. Sun, T.-Y. Yun, A low-power and high-gain mixer for UWB systems. IEEE Microwave Wirel. Compon. Lett. **18**(12), 803–805 (2008)
29. A.V. Do, C.C. Boon, M.A. Do, K.S. Yeo, A. Cabuk, A weak-inversion low-power active mixer for 2.4 GHz ISM band applications. IEEE Microwave Wirel. Compon. Lett. **19**(11), 719–721 (2009)
30. P. Andricciola, H.P. Tuinhout, The temperature dependence of mismatch in deep-submicrometer bulk MOSFETs. IEEE Electron Device Lett. **30**(6), 690–692 (2009)

Chapter 7
Summary and Conclusions

Abstract Summarizing remarks and conclusions are provided in this chapter with regards to system-level self-calibration strategies as well as the specific case studies described in the book.

7.1 Overall Perspective

Contemporary CMOS technologies make it possible to design highly integrated multi-functional chips. On the other hand, the current research and product development trends are associated with several challenges in the quality assurance and reliability of the manufactured devices. As described in Chap. 1, many problems are fundamentally caused by worsening process parameter variations, interactions between individual blocks through coupling effects on the same chip, system complexity, high on-chip power densities, and the increasing number of functions to be verified. In the case of wireless systems, an additional issue is that more and more devices are designed to transmit/receive signals from multiple communication standards, leading to interference problems. A survey of the existing and emerging on-chip built-in test and calibration techniques for single-chip wireless transceivers was presented in this book. Since it embodies various design philosophies in academia and the industry, the overview exposed the diversity among the approaches to solve the current testability and reliability challenges. In general, it can be observed that a tendency exists to combine system-level test and calibration techniques with digitally adaptable circuits within the analog sections of the transceivers, where the digital processor monitors system parameters and controls corrective actions.

   Supplemental measurements or calibration loops on the analog circuit level are beneficial to quickly detect and correct gross variations at start-up in order to

M. Onabajo and J. Silva-Martinez, *Analog Circuit Design for Process*   151
Variation-Resilient Systems-on-a-Chip, DOI: 10.1007/978-1-4614-2296-9_7,
© Springer Science+Business Media New York 2012

reduce the computational overhead and time requirements in the digital processor. On-chip built-in test circuitry also aids the identification of fault location to determine appropriate adjustments. Moreover, certain faults are extremely difficult to observe in the digital baseband of receivers, particularly defects and variations in the RF front-end section such as those related to impedance matching. Hence, many on-chip built-in test and calibration techniques involve analog measurement circuitry. The emphasis in this book was on the exploration of design strategies to make analog circuits more robust to PVT variations. Since this task is very specific to the type of circuit being designed, several examples with different analog and mixed-signal circuits in wireless receivers were discussed. In general, it can be concluded that variation-aware analog design itself is not sufficient to guarantee the required performance in demanding applications. For this reason, it is advisable to equip the analog blocks with features for performance tuning during production testing or even during normal operation of the devices. Most of the alterations of the example circuits in this book encompass digitally programmable elements for compatibility with the system-level calibration approaches that were addressed in Chap. 2.

7.2 Discussed Projects

The first example involved the design task to increase the linearity of operational transconductance amplifiers (OTAs) in lowpass filters with wide bandwidth. In Chap. 3, an architectural solution was discussed which is based on cancellation of the main amplifier's nonlinearities with an identical auxiliary OTA. With regards to resilience to PVT variations, the motivation for this approach is that two amplifiers with the same component dimensions and bias conditions exhibit minimal mismatches. This characteristic is particularly important to arrive at an effective broadband linearization method because it ensures minimal deviations of the high-frequency responses in the main and auxiliary signal paths. Nevertheless, the analysis of the problem and experimental results have revealed that high linearity at high frequencies requires the ability to compensate for PVT variations. To do so, digitally programmable resistor ladders were utilized to perform the necessary post-fabrication gain and phase equalizations for optimum cancellation of nonlinearities. Measurements obtained with a 0.13 µm CMOS test chip demonstrated that the nonlinearity cancellation technique improves the IM3 of the designed OTA by up to 22 dB at frequencies up to 350 MHz. Consuming 5.2 mW from a 1.2 V supply, the linearized OTA with a $0.2V_{p-p}$ input signal has an IM3 better than −74 dB up to 350 MHz and a 70 dB signal-to-noise ratio (SNR) in 1 MHz bandwidth. The linearization scheme was also tested with multiple OTAs embedded into a lowpass filter having a 195 MHz bandwidth. This filter has a measured in-band IIP3 of 14.0 dBm and a 54.5 dB dynamic range.

In the second presented circuit example, the quantizer topology in Chap. 4 was developed as part of a continuous-time $\Sigma\Delta$ modulator architecture with 3-bit pulse-width modulation in the feedback path in order to circumvent the nonlinearity problems caused by unit element mismatches in multi-bit feedback circuitry. Besides robustness to process variations, the other incentives for using this $\Sigma\Delta$ modulator architecture are the scalability and the potential for power savings with state of the art CMOS technology. However, low-jitter clocks are required for this time-based architecture, which is why the 7-phase 400 MHz clock signal is provided by an injected-locked clock generator. A two-step current-mode quantizer was designed for the $\Sigma\Delta$ modulator. This 3-bit quantizer utilizes the available clock phases for analog-to-digital conversion with successive approximations. If applications require tuning for finer resolution, the high-impedance of the reference voltage inputs allow them to be generated with low-power on-chip digital-to-analog converters as those used in many system-level calibration schemes. The quantizer functionality was verified through the measurements of the 5th-order continuous-time $\Sigma\Delta$ modulator chip with the embedded quantizer, which was fabricated in a 0.18 μm CMOS process.

Better observability of faults and variations usually improves the accuracy or execution time of test and calibration routines, for which electrical detectors and process monitoring circuits are utilized. Towards this end, a temperature sensing approach has been assessed in Chap. 5. Since this alternative technique does not require a connection to the circuit under test or signal path, it provides a non-influential method for monitoring variations. A design procedure with electro-thermal co-simulation was outlined to evaluate RF circuit performance metrics from the DC output of an on-chip temperature sensor. The described fully-differential sensor circuit for this application has been designed with a wide dynamic range, programmable sensitivity to DC and RF power dissipation, and compatibility with CMOS technology. Using an LNA as prototype, measurements obtained with a 0.18 μm CMOS technology test chip showed that RF power dissipation can be observed with the on-chip temperature sensor. Furthermore, the 1 dB compression point can be estimated with less than 1 dB error. The sensor circuitry with 0.012 mm^2 die area can be shared when several on-chip test points are monitored by placement of multiple temperature-sensing parasitic bipolar devices having an emitter area of 11×11 μm.

Finally, an alternative approach to alleviating the effects of process parameter variations was explained in Chap. 6. Rather than employing digitally adjustable elements, the mismatch reduction scheme employs an automatic analog calibration loop to improve the matching of transistors in the high-frequency differential signal path. The method is intended for analog circuits in which short-channel devices are used to minimize bandwidth reduction from parasitic capacitances, and in which transistors are not directly matched to reduce high-frequency coupling through layout parasitics and substrate leakage. Monte Carlo simulations were performed to evaluate the approach for two example circuits designed with 90 nm and 0.13 μm CMOS technology. In the first case, the application of the mismatch reduction loop to a differential amplifier with 13 dB gain and a −3 dB frequency of 2.14 GHz lowered the simulated standard deviation of the input-referred offset voltage from 4.17 to 0.76–1.29 mV, depending on the assumed layout of the sensing-transistors. In the second case, the mismatch reduction loop was used to boost the simulated IIP2 of a double-balanced mixer by 5 dB via improvement of the matching between the switching transistors. Based on the results, it can be concluded that this mismatch reduction scheme is suitable for fast coarse calibration at start-up because the loop's settling time can be kept in the range of a few microseconds. If further calibration accuracy is needed and on-chip digital resources are available, then it could be explored to merge the analog loop with digitally-controlled elements within the mixer for system-level calibration with longer convergence.

Appendix A
OTA Linearization—Volterra Series Analysis

In this appendix, the optimum compensation resistor value for linearization at high frequencies is derived with Volterra series analysis [1]. Employing a third-order model of transconductor nonlinearity, the simplified model of the attenuation-predistortion linearization technique is shown in Fig. A.1. In this analysis, g_{m1} represents the linear transconductance and g_{m3} the third-order component. Resistor (R_c) compensates for high-frequency linearity degradation by equalizing the delays in the main and auxiliary paths. The differential voltage $V_{i2}(t)$ at the input of the main OTA is given by

$$V_{i2}(t) = -\left(g_{m1}k_2 V_{in}(t) + g_{m3}[k_2 V_{in}(t)]^3 \right) \cdot \frac{R \cdot (1 - k_1)}{1 + 2C_p/C} \cdot \frac{1 + j\omega C k_1 R_c}{1 + j\omega b - c\omega^2}$$

$$+ V_{in}(t) \cdot \frac{k_1}{1 + 2C_p/C} \cdot \frac{1 + j\omega C(1 - k_1)R/2 + j\omega C_o R}{1 + j\omega b - c\omega^2} \quad \text{(A.1a)}$$

where:

$$b = \frac{C(k_1/2)(1 - k_1)(R + 2R_c) + 2k_1 C_p R_c + (1 - k_1)C_p R}{1 + 2C_p/C} + C_o R$$

$$c = \frac{k_1(1 - k_1)CC_p RR_c + CC_o k_1(1 - k_1)RR_c + 2k_1 C_p C_o R_c R}{1 + 2C_p/C} \quad \text{(A.1b)}$$

Following the same analysis as in Sect. 3.2.3 but taking the parasitic capacitances C_p and C_o into account, the conditions for distortion cancellation at low frequencies are:

$$\frac{g_{m1} \cdot R \cdot (1 - k_1)}{1 + 2C_p/C} = 1, \quad k_2 = \frac{k_1/2}{1 + 2C_p/C} \quad \text{(A.2)}$$

M. Onabajo and J. Silva-Martinez, *Analog Circuit Design for Process Variation-Resilient Systems-on-a-Chip*, DOI: 10.1007/978-1-4614-2296-9,
© Springer Science+Business Media New York 2012

Fig. A.1 Nonlinear model
for differential attenuation-
predistortion cancellation

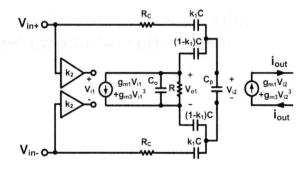

With the above provisions, the output current of the main OTA after algebraic simplifications is:

$$
\begin{aligned}
i_{out}(t) &= g_{m1}V_{i2}(t) + g_{m3}[V_{i2}(t)]^3 \\
&\approx g_{m1}V_{in}(t) \cdot \frac{(k_1/2)}{1 + 2C_p/C} \cdot \frac{1 + j\omega C((1 - k_1)R - k_1R_c) + j2\omega C_o R}{1 + j\omega b - c\omega^2} \\
&\quad - g_{m3}\left(\frac{k_1 V_{in}(t)/2}{1 + 2C_p/C}\right)^3 \cdot \frac{1 + j\omega C k_1 R_c}{1 + j\omega b - c\omega^2} \\
&\quad + g_{m3}\left(\frac{k_1 V_{in}(t)/2}{1 + 2C_p/C} \cdot \frac{1 + j\omega C((1 - k_1)R - k_1R_c) + j2\omega C_o R}{1 + j\omega b - c\omega^2}\right)^3
\end{aligned} \tag{A.3}
$$

Assuming weakly nonlinear operation based on condition (3) in Sect. 3.2.3 and that the signal can be expressed as a sum of sinusoids with incommensurate frequencies, the harmonic input method can be applied to calculate the Volterra series coefficients [1] and theoretically demonstrate the nonlinearity cancellation with the attenuation-predistortion linearization scheme. Taking a single input $V_{in}(t) = exp(j\omega_1 t)$ and substituting into (A.3) to express the linear transfer function H_1:

$$
H_1 = g_{m1} \cdot \frac{(k_1/2)}{1 + 2C_p/C} \cdot \frac{1 + j\omega C((1 - k_1)R - k_1R_c) + 2j\omega C_o R}{1 + j\omega b - c\omega^2} \tag{A.4}
$$

Selecting $V_{in}(t) = exp(j\omega_1 t) + exp(j\omega_2 t) + exp(j\omega_3 t)$ and making the appropriate substitutions for calculation of the third-order transfer function (H_3) yields the following equality after expansion and omission of all terms that do not contain the $exp(j\omega_1 t + j\omega_2 t + j\omega_3 t)$ factor relevant to H_3:

$$H_3(\omega_1,\omega_2,\omega_3) = g_{m3} \left(\frac{k_1/2}{1+2C_p/C} \right)^3$$

$$\times \left(\frac{1+j\omega_1 C((1-k_1)R-k_1 R_c)+2j\omega_1 C_o R}{1+j\omega_1 b-c\omega_1^2} \right)$$

$$\times \left(\frac{1+j\omega_2 C((1-k_1)R-k_1 R_c)+2j\omega_2 C_o R}{1+j\omega_2 b-c\omega_2^2} \right)$$

$$\times \left(\frac{1+j\omega_3 C((1-k_1)R-k_1 R_c)+2j\omega_3 C_o R}{1+j\omega_3 b-c\omega_3^2} \right)$$

$$-g_{m3} \left(\frac{k_1/2}{1+2C_p/C} \right) \frac{1+j(\omega_1+\omega_2+\omega_3)Ck_1 R_c}{1+j(\omega_1+\omega_2+\omega_3)b-c(\omega_1+\omega_2+\omega_3)^2} \quad (A.5)$$

The amplitude of the third harmonic distortion (HD3) current due to a sinusoidal input signal $V_{in} \sin(\omega t)$ is given by

$$i_{o3} = \frac{1}{4} V_{in}^3 H_3(\omega, \omega, \omega)$$

$$= \frac{1}{4} g_{m3} \left(\frac{V_{in} k_1/2}{1+2C_p/C} \right)^3 \left(\frac{1+j\omega c((1-k_1)R-k_1 R_c)+2j\omega C_o R}{1+j\omega b-c\omega^2} \right)^3$$

$$-\frac{1}{4} g_{m3} \left(\frac{V_{in} k_1/2}{1+2C_p/C} \right)^3 \frac{1+j3\omega Ck_1 R_c}{1+j3\omega b-9c\omega^2} \quad (A.6)$$

Elimination of HD3 requires that $i_{o3} = 0$, hence

$$\frac{1+j\omega C((1-k_1)R-k_1 R_c)+2j\omega C_o R}{1+j\omega b-c\omega^2} = \sqrt[3]{\frac{1+j3\omega Ck_1 R_c}{1+j3\omega b-9c\omega^2}} \quad (A.7)$$

The cubic root in (A.7) can be approximated with $(1+x)^{1/3} \approx 1 + x/3$ for $x \ll 1$. Thus,

$$\frac{1+j\omega C((1-k_1)R-k_1 R_c)+2j\omega C_o R}{1+j\omega b-c\omega^2} \approx \frac{1+j\omega Ck_1 R_c}{1+j\omega b-3c\omega^2}$$

$$\Rightarrow R_c \approx \frac{(1-k_1)+2C_o/C}{2k_1} R \quad \text{to cancel HD3} \quad (A.8)$$

For a two-tone input signal of the form $V_{in1} \sin(\omega_1 t) + V_{in2} \sin(\omega_2 t)$, the IM3 current can be determined with Volterra series [1] according to the following equation:

$$i_{IM3} = \frac{3}{4} V_{in1}^2 V_{in2} H_3(\omega_1, \omega_1, -\omega_2)$$

$$= g_{m3} \left(\frac{k_1/2}{1+2C_p/C} \right)^3 (3V_{in1}^2 V_{in2}/4) \left(\frac{1+j\omega_1 C((1-k_1)R - k_1 R_c) + 2j\omega_1 C_o R}{1+j\omega_1 b - c\omega_1^2} \right)^2$$

$$\times \left(\frac{1+j\omega_2 C((1-k_1)R - k_1 R_c) + 2j\omega_2 C_o R}{1+j\omega_2 b - c\omega_2^2} \right)$$

$$- g_{m3} \left(\frac{k_1/2}{1+2C_p/C} \right)^3 (3V_{in1}^2 V_{in2}/4) \frac{1+j(2\omega_1 - \omega_2)Ck_1 R_c}{1+j(2\omega_1 - \omega_2)b - c(2\omega_1 - \omega_2)^2} \qquad \text{(A.9)}$$

Simplifying i_{IM3} for two intermodulation tones that are close together $(\omega_1 \approx \omega_2 \approx 2\omega_1 - \omega_2)$ yields:

$$i_{IM3} \approx g_{m3} \left(\frac{k_1/2}{1+2C_p/C} \right)^3 (3V_{in1}^2 V_{in2}/4) \left(\frac{1+j\omega_1 C((1-k_1)R - k_1 R_c) + 2j\omega_1 C_o R}{1+j\omega_1 b - c\omega_1^2} \right)^2$$

$$\times \left(\frac{1-j\omega_1 C((1-k_1)R - k_1 R_c) + 2j\omega_1 C_o R}{1+j\omega_1 b - c\omega_1^2} \right)$$

$$- g_{m3} \left(\frac{k_1/2}{1+2C_p/C} \right)^3 (3V_{in1}^2 V_{in2}/4) \frac{1+j\omega_1 Ck_1 R_c}{1+j\omega_1 b - c\omega_1^2}$$

$$\Rightarrow R_c \approx \frac{(1-k_1)+2C_o/C}{2k_1} R \quad \text{for } i_{IM3} \approx 0 \qquad \text{(A.10)}$$

Appendix B
OTA Linearization—Advanced Phase Compensation

Figure B.1a depicts a model for an OTA in integrator configuration where r_o represents the OTA output impedance and $Gm_{(j\omega)}$ the transconductance that changes with frequency due to internal parasitic poles. Both nonidealities cause deviations from ideal integration on the load capacitor C. The following analysis shows that the linearization introduces an additional pole, which can be cancelled by adding resistor R_s in series with the load capacitor as in the conventional case [2].

Let $\omega_o = 1/(r_oC)$ be the dominant pole of the integrator configuration and ω_1 be the internal parasitic pole of the OTA with the lowest frequency. If $\omega_1 \gg \omega_o$, then the transfer function of the configuration in Fig. B.2a is:

$$\frac{V_{o+} - V_{o-}}{V_{i+} - V_{i-}} = \frac{Gm_{(0)} \cdot r_o}{1 + s \cdot r_oC} \approx \frac{Gm_{(0)}}{s \cdot C} \tag{B.1}$$

where $s = j\omega$ and the approximation implies: $\omega_o \ll \omega \ll \omega_1$. When using attenuation-predistortion linearization at high frequencies, the additional pole ω_c formed by R_c and C_p in Fig. 3.2 is not negligible in all designs. Hence, the integrator has the following transfer function:

$$\frac{V_{o+} - V_{o-}}{V_{i+} - V_{i-}} = \frac{Gm_{(0)} \cdot r_o}{1 + s \cdot r_oC} \cdot \frac{1}{1 + s/\omega_c} \tag{B.2}$$

To avoid impact of ω_c, a series resistor R_s can be added to the load capacitor C as visualized in Fig. B.1b, resulting in the new expression for the transfer function:

$$\frac{V_{o+} - V_{o-}}{V_{i+} - V_{i-}} = \frac{Gm_{(0)} \cdot r_o(1 + s \cdot R_sC)}{(1 + sC[r_o + R_s]) \cdot (1 + s/\omega_c)} \approx \frac{Gm_{(0)} \cdot r_0(1 + s \cdot R_sC)}{(1 + s \cdot r_oC) \cdot (1 + s/\omega_c)} \tag{B.3}$$

where $R_s \ll r_o$ is assumed in the approximation. In the range $\omega_o \ll \omega \ll \omega_1$, the following condition to nullify the impact of the linearization can be identified by comparing (B.3) and (B.1):

M. Onabajo and J. Silva-Martinez, *Analog Circuit Design for Process Variation-Resilient Systems-on-a-Chip*, DOI: 10.1007/978-1-4614-2296-9,
© Springer Science+Business Media New York 2012

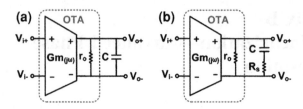

Fig. B.1 OTA model with additional nonidealities: **a** Standard configuration, **b** configuration to compensate for the internal pole from the attenuation-predistortion linearization; where $Gm_{(\omega)}$, r_o, and C are the frequency-dependent transconductance, finite output impedance, and load capacitor, respectively

Fig. B.2 Single-ended equivalent block diagram of a bandpass biquad

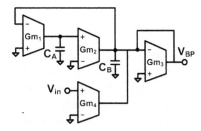

$$R_s = \frac{1}{C \cdot \omega_c} \tag{B.4}$$

The effect of ω_c from the linearization on the key parameters of a biquad section can be assessed by examining the bandpass (BP) case (Fig. B.2). The center frequency (ω_{oi}), bandwidth (BW_i), and quality factor (Q_i) with ideal OTAs are:

$$\omega_{oi} = \sqrt{\frac{Gm_1 Gm_2}{C_A C_B}} \tag{B.5}$$

$$BW_i = \frac{Gm_3}{C_B} \tag{B.6}$$

$$Q_i = \sqrt{\frac{Gm_1 Gm_2}{(Gm_3)^2} \cdot \frac{C_B}{C_A}} \tag{B.7}$$

Substituting $Gm_{(s)} = Gm/(1 + s/\omega_c)$ for each Gm in the BP transfer function yields the following equation for a linearized BP section:

$$H_{BP}(s) = \frac{V_{BP}}{V_{in}} = \frac{N(s)}{D(s)}$$

$$= \frac{\dfrac{Gm_4}{C_B} \cdot \dfrac{1}{1 + s/\omega_{c4}} \cdot s}{s^2 + \dfrac{Gm_3}{C_B} \cdot \dfrac{1}{1 + s/\omega_{c3}} \cdot s + \dfrac{Gm_1 Gm_2}{C_A C_B} \cdot \dfrac{1}{1 + s/\omega_{c1}} \cdot \dfrac{1}{1 + s/\omega_{c2}}} \tag{B.8}$$

Fig. B.3 Single-ended diagram of a bandpass biquad with phase compensation

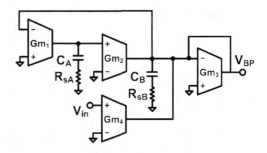

Letting $Gm = Gm_1 = Gm_2$ and $\omega_c = \omega_{c1} = \omega_{c2}$ for simplicity and given that $\omega_{oi} < \omega_c$, it can be shown that the center frequency (ω_{on}) of the linearized BP biquad can be approximated as:

$$\omega_{on} \approx \omega_{oi} \tag{B.9}$$

The denominator of the linearized BP transfer function in (B.8) can be approximated as follows:

$$
\begin{aligned}
D(s) &\approx s^2 + \frac{Gm_3}{C_B} \cdot (1 - s/\omega_{c3}) \cdot s + \frac{Gm^2}{C_A C_B} \cdot (1 - 2s/\omega_c) \\
&\approx s^2 + \left(\frac{Gm_3}{C_B} - \frac{2Gm^2}{\omega_c C_A C_B} \right) \cdot s + \frac{Gm^2}{C_A C_B}
\end{aligned}
\tag{B.10}
$$

where the second approximation is valid when $\omega \ll \omega_{c3}$. From (B.10), BW_n and the quality factor (Q_n) with linearized OTAs can be written in terms of the above ideal expressions as follows:

$$BW_n \approx \frac{Gm_3}{C_B} - \frac{2Gm^2}{\omega_c C_A C_B} = BW_i - \frac{2\omega_{oi}^2}{\omega_c} = BW_i \left(1 - \frac{2\omega_{oi}^2}{\omega_c \cdot BW_i} \right) \tag{B.11}$$

$$Q_n = \frac{\omega_{on}}{BW_n} \approx \frac{\omega_{oi}}{BW_i \left(1 - \dfrac{2\omega_{oi}^2}{\omega_c \cdot BW_i} \right)} \tag{B.12}$$

Equation (B.12) shows that the quality factor error from linearization increases with the ratio of $\omega_{oi}^2 / (\omega_c \approx BW_i)$, where: $\omega_{oi} \approx \omega_{on}$. Furthermore, stability requires:

$$\frac{2\omega_{oi}^2}{\omega_c \cdot BW_i} < 1 \tag{B.13}$$

The parameter changes in (B.11, B.12) can be incorporated into the design of linearized biquads by altering the transconductance and capacitor values accordingly. Alternatively, the effects of the linearization can be canceled as described next.

Using series resistors R_{sA} and R_{sB} with C_A and C_B to compensate for the phase shift from the linearization as described above and shown in Fig. B.3, the corresponding zeros are introduced in the denominator:

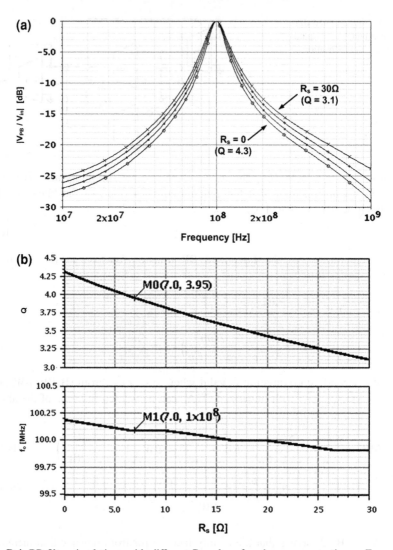

Fig. B.4 BP filter simulations with different R_s values for phase compensation. **a** Frequency responses, **b** quality factor and center frequency; where $R_s = R_{sA} = R_{sB} \cdot (C_B/C_A)$

$$D(s) = s^2 + \frac{Gm_3}{C_B} \cdot \frac{1 + s/\omega_{zB}}{1 + s/\omega_{c3}} \cdot s + \frac{Gm^2}{C_A C_B} \cdot \frac{(1 + s/\omega_{zA})(1 + s/\omega_{zB})}{(1 + s/\omega_c)^2} \quad \text{(B.14)}$$

where $\omega_{zA} = 1/(R_{sA}C_A) = \omega_{zB} = 1/(R_{sB}C_B) = \omega_z = \omega_c$. Using the same approximations as in (B.9–B.13), the compensated center frequency (ω_{cn}) and bandwidth (BW_{cn}) become:

$$\omega_{cn} \approx \sqrt{\frac{Gm^2}{C_A C_B} \cdot (1 - 2s/\omega_c) \cdot (1 + 2s/\omega_z)} \approx \sqrt{\frac{Gm^2}{C_A C_B} \cdot \left(1 - \frac{4s^2}{\omega_c \cdot \omega_z}\right)} \approx \sqrt{\frac{Gm^2}{C_A C_B}}$$

$$(B.15)$$

which is equivalent to ω_i as a result of the last simplification step ($4 \cdot \omega^2 \ll \omega_c \cdot \omega_z$).

$$BW_{cn} \approx \frac{Gm_3}{C_B} \cdot \left[1 + \left(\frac{1}{\omega_z} - \frac{1}{\omega_{c3}}\right) \cdot s - \frac{s^2}{\omega_z \cdot \omega_{c3}}\right] \approx \frac{Gm_3}{C_B} \cdot \left[1 + j\omega \cdot \left(\frac{1}{\omega_z} - \frac{1}{\omega_{c3}}\right)\right]$$

$$(B.16)$$

Note, a small bandwidth error remains after compensation due to the difference between ω_z and ω_{c3} because $\omega_{zA}(R_{sA}, C_A)$ and $\omega_{zB}(R_{sB}, C_B)$ are optimized to cancel ω_{c1} and ω_{c2} of Gm_1 and Gm_2, respectively. Thus, the pole ω_{c3} is only partially cancelled if $Gm_1 \neq Gm_3$. Nevertheless, the second term in (B.16) has a small effect in the typical case and $BW_{cn} \approx BW_i$ since $\omega \ll |1/\omega_z - 1/\omega_{c3}|^{-1}$.

The linearized OTAs described in Sect. 3.3.1 were employed in a BP filter (Fig. B.3) with $f_o = 100$ MHz, $Gm_3 = Gm_4 = Gm/2$, and $Gm = Gm_1 = Gm_2$ for simplicity (implying $\omega_c = \omega_{c1} = \omega_{c2}$). Series resistors R_{sA} and R_{sB} with C_A and C_B compensate for the phase shift from the linearization by creating zeros ω_{zA} and ω_{zB}: $\omega_{zA} = 1/(R_{sA}C_A) = \omega_{zB} = 1/(R_{sB}C_B) = \omega_z = \omega_c$. A small BW error remains after compensation due to the difference between ω_z and ω_{c3} of Gm_3 because $\omega_{zA}(R_{sA}, C_A)$ and $\omega_{zB}(R_{sB}, C_B)$ are optimized to cancel ω_{c1} and ω_{c2} of Gm_1 and Gm_2, respectively. Thus, the pole ω_{c3} is only partially cancelled since $Gm_1 \neq Gm_3$. Nevertheless, the effect is small in the typical case ($\omega \ll \omega_{c3}$). This BP filter achieves simulated IM3 of -72.0 dB evaluated after an additional output buffer (Gm). Figure B.4 contains simulated plots of the frequency responses for different values of R_s from this example BP filter design. The plots show how the adjustment of $R_s = R_{sA} = R_{sB} \cdot (C_B/C_A)$ during the design allows tuning of the quality factor to ~ 4 with $R_s = 7\,\Omega$ in this case, while f_o does not change significantly.

Appendix C
OTA Linearization Without Power Budget Increase

Attenuation-predistortion linearization offers the means to improve the linearity of a given OTA while preserving its AC characteristics without design changes in the OTA core, which is achieved at the expense of increased power, noise, and layout area. Another option is to redesign the two OTAs in the linearization scheme using half of the power in order to meet the same power budget as the original OTA. But, that approach is associated with a reduction of the OTA bandwidth as delineated in this appendix.

To accomplish linearization with equal power budget, the currents I_b and I_{b1} in Fig. 3.4 can be reduced by 50%, which requires increasing the W/L ratios of the transistors in the core (M_c) to obtain the same transconductance as before. Thus, the saturation voltage V_{DSAT} of M_c becomes approximately half of the initial value. Furthermore, the ratio of transconductance to parasitic capacitance (i.e. f_T) of both OTAs in the linearization scheme reduces due to the bias current decrease and width increase for M_c. Gain vs. frequency simulations of the linearized OTA (50% power reduction in each path) and the reference OTA revealed that the linearization with equal power reduces the effective 3 dB bandwidth from 2.49 to 1.09 GHz with 50 Ω load. Table C.1 summarizes the key results from simulating the linearized OTA in comparison to the reference OTA with identical total power. High linearity through distortion cancellation (IM3 $\approx$ −77 dB) is achievable, but limited to lower frequencies. Despite of this, the results indicate that higher FOM (see Table 3.3) can be achieved with low-frequency linearization compared to the linearization with doubled power consumption.

M. Onabajo and J. Silva-Martinez, *Analog Circuit Design for Process Variation-Resilient Systems-on-a-Chip*, DOI: 10.1007/978-1-4614-2296-9, © Springer Science+Business Media New York 2012

Table C.1 Simulated comparison: OTA linearization without power increase

OTA type	V_{DSAT} of input differential pair (M_c) (mV)	f_{3db} with 50 Ω load (GHz)	Input-referred noise (nV/$\sqrt{Hz}$)	Power (mW)	IM3 ($V_{in} = 0.2\ V_{p-p}$)	Normalized |FOM|[a] (at f_{max})
Reference (input attenuation = 1/3)	90	2.49	9.7	2.6	−53.1 dB at f_{max} = 350 MHz (−53.2 dB at 100 MHz)	57.2
Linearized (attenuation = 1/3 & compensation)	54	1.09	14.3	2.6	−77.1 dB at f_{max} = 100 MHz	119.2

[a] See Table 3.3 for details

Appendix D
Temperature Sensing Analysis—Relationship Between Circuit Nonlinearities and DC Temperature

The main purpose of the analysis in this appendix is to show that a minimum temperature point is sensed near a MOS device as the RF power of an applied signal is swept. When a sinusoidal input voltage $x(t)$ with amplitude X at frequency ω excites a weakly nonlinear MOS device and creates a current $y(t)$ that can be expressed by a power series with coefficients α_0, α_1, α_2,...; then the signals can be written as

$$x(t) = X \cos \omega t \tag{D.1}$$

$$y(t) = \alpha_0 + \alpha_1 x(t) + \alpha_2 x^2(t) + \alpha_3 x^3(t) + \dots \tag{D.2}$$

The effect of the bias current α_0 is removed via calibration before the application of the signal, which avoids interference with the 1-dB compression point characterization. Thus, the signal-dependent current without α_0 can be expressed as

$$y_{sig}(t) = y_{sig}(t)|_{DC} + y_{sig}(t)|_{AC} \tag{D.3}$$

where:

$$y_{sig}(t)|_{DC} \approx \frac{\alpha_2}{2} X^2 \tag{D.4}$$

$$y_{sig}(t)|_{AC} = \left(\alpha_1 X + \frac{3\alpha_3}{4} X^3\right) \cos \omega t$$
$$+ \frac{\alpha_2}{2} X^2 \cos 2\omega t + \frac{\alpha_3}{4} X^3 \cos 3\omega t + \dots \tag{D.5}$$

A conventional 1-dB compression point characterization is a measure of the third-order distortion due to α_3 at frequency ω, for which the input amplitude approximation is given by

M. Onabajo and J. Silva-Martinez, *Analog Circuit Design for Process Variation-Resilient Systems-on-a-Chip*, DOI: 10.1007/978-1-4614-2296-9,
© Springer Science+Business Media New York 2012

$$X_{1dB} = \sqrt{\left|(4/3) \cdot (10^{-1/20} - 1)\frac{\alpha_1}{\alpha_3}\right|} \tag{D.6}$$

With the homodyne temperature sensing approach, the linearity is assessed from indirect measurement of the DC power, giving rise to the implications analyzed below.

When a signal is applied, the AC amplitude and the signal-dependent part of the drain-source voltage's DC component resulting from $y_{sig}(t)|_{DC}$ scale proportionally to the RMS drain-source voltage change. Let K represent this load-dependent proportionality factor. In the transistors of the CUT, the AC drain-source voltage is 180° out of phase with the drain current $y_{sig}(t)|_{AC}$. Thus, a simplified approximation for the signal-dependent drain-source voltage around the 1-dB compression point is:

$$\begin{aligned} v_{sig[1dB]}(t) &= K \cdot y_{sig}(t)\big|_{DC[1dB]} - K \cdot y_{sig}(t)\big|_{AC[1dB]} \\ &\approx K_{DC} - K_{AC} \cdot \cos(\omega t) \end{aligned} \tag{D.7}$$

where:

$$K_{DC} = K \cdot \frac{\alpha_2}{2} X_{1dB}^2 \tag{D.8}$$

$$K_{AC} = K \cdot \left(\alpha_1 X_{1dB} + (3/4) \cdot \alpha_3 X_{1dB}^3\right) \tag{D.9}$$

Here, the analysis is simplified by the omission of load-dependent nonlinearities and by disregarding components at 2ω, 3ω, and higher harmonics. More complex expressions and incorporation of electro-thermal coupling would be required for more accurate analytical estimates. Nevertheless, the approximations under the assumed conditions give insights into the key characteristics of the power that causes the temperature change:

$$p(t) = v_{sig[1dB]}(t) \cdot y_{sig}(t) \tag{D.10}$$

Notice that (D.10) only represents the scaled signal-dependent power components. Without calibration, the DC bias current (α_0) would have to be included and multiplied with a different factor (unrelated to K). But, α_0 was dropped from (D.2) because its contribution is nullified by the calibration step. After substituting (D.3–D.5) and (D.7–D.9) into (D.10), using the trigonometric identity $\cos^2(x) = \frac{1}{2} \cdot [1 + \cos(2x)]$, and dropping all remaining AC terms based on the low-pass filter characteristics of the thermal coupling (under condition: $\omega \gg 2\pi \cdot 10$ kHz); the DC power component that causes the measured DC temperature change is obtained as follows:

$$P_{DC \to \Delta T} \approx K_{DC} \cdot \left(\frac{\alpha_2}{2} X^2\right) - (1/2) \cdot K_{AC} \cdot \left(\alpha_1 X - (3/4) \cdot |\alpha_3| X^3\right) \tag{D.11}$$

The approximation in (D.11) assumes weakly nonlinear operation, negligible higher-order distortion components, and the typical case in which α_0–α_2 are positive but α_3 is negative. From (D.11), it can be observed that second-order nonlinearity creates a measurement offset and that the DC component reaches a minimum as X is swept to evaluate the 1-dB compression property. This theoretical minimum can be derived by taking the derivative of (D.11), equating the resulting expression to zero, and solving for the amplitude:

$$X_{\min} = \frac{-\alpha_2(K_{DC}/K_{AC}) + \sqrt{\alpha_2^2(K_{DC}/K_{AC})^2 + (9/4) \cdot \alpha_1|\alpha_3|}}{(9/4) \cdot |\alpha_3|} \tag{D.12}$$

Equation (D.12) gives insights into the minimum temperature point characteristics, but it is important to note that it is only a rough approximation due to the aforementioned assumptions. In the absence of thermal coupling to other devices, a relative comparison of (D.12) and (D.6) allows to estimate the fixed input power shift (in decibels) between the minimum DC power/temperature point and the 1 dB-compression point:

$$shift_{\min[1dB]} = 10 \cdot \log(X_{\min}^2/X_{1dB}^2) \tag{D.13}$$

The above equations show that the 1-dB compression point can be inferred from the DC power dissipation monitored by the temperature sensor as long as the second-order nonlinearity is accounted for during simulations. For the nonlinearity coefficients of the example CUT, (D.13) predicts a 4.73 dB shift. Based on this shift with respect to the simulated 0.5 dBm 1-dB compression point, the minimum point is expected with 5.23 dBm input power. However, P_m in Fig. 5.5 has a minimum at 2.6 dBm, where the error is mainly caused by the aforementioned idealizations and by deviations from the weak nonlinearity assumption that causes approximately 15% error in (D.6) alone. Furthermore, the thermal coupling of devices in the CUT affects the minimum temperature point on the x-axis, which follows the superposition principle (e.g. the power of all devices in Fig. 5.5 results in the combined temperature curve (T_s) at the sensing PNP device in Fig. 5.6). Therefore, the electro-thermal simulation method presented in this book provides a more reliable estimate for the shift, which was around 0.1 dB in simulations and 0.5 dB in measurements. The difference is affected by process variations as well as electro-thermal modeling inaccuracies, which could cause up to ±0.6 dB uncertainty for this CUT that was added to the measurement error.

References

1. S. Maas, *Nonlinear Microwave and RF Circuits* (Artech House, Boston, 2003)
2. A.P. Nedungadi, R.L. Geiger, High-frequency voltage-controlled continuous time lowpass filter using linearized CMOS integrators. Electron. Lett. **22**, 729–731 (1986)

Index

1

1-dB compression point, 18, 89, 94, 103, 109–111, 136

A

Amplitude detectors, 17
Analog calibration, 16, 113, 114, 123
Analog front-end, 14, 16, 17
Analog measurements, 13, 20
Analog tuning, 15, 16, 21
Analog-to-digital converter, 14, 32, 57, 89
Attenuation-predistortion linearization, 32, 35, 38, 41, 51
Automatic test equipment, 21

B

Baseband, 14, 17–19, 24, 25, 31, 41, 48, 58, 123
Bit error rate, 13, 14, 41
Block-level measurement, 17
Built-in self-test, 10
Built-in test, 13, 21, 87, 111, 112

C

Calibration, 13, 41, 89, 106, 113
Characterization, 9, 17, 88, 101, 105, 111, 112
Circuit under test, 15, 17, 87, 88, 111
Common-mode feedback, 40, 99, 122, 125, 131, 132
Communication standards, 12
Comparator, 42, 62, 67, 68, 70, 72–75, 84, 85

Continuous-time, 57, 75, 85
Control voltage, 25, 40
Convergence time, 16, 114, 124
Cost of testing, 10

D

DC bias, 25, 33, 94, 97, 106
DC offsets, 14, 19, 24, 123
Defect-oriented test, 21
Design-for-manufacturability, 10
Design-for-test, 10
Device mismatches. See mismatches, 63
Differential temperature sensor, 87, 88, 103
Digital calibration, 14, 60, 123, 148
Digital correction, 13, 14, 24, 31, 32, 41, 44
Digital signal processor, 19, 41, 89, 123
Digitally-controlled, 25, 41
Digital-to-analog converter, 19, 57, 58, 60
Distortion cancellation, 32–35, 42, 43, 45
Dynamic range, 37, 51, 80, 85, 87, 94, 96–98, 100, 102–104, 111

E

Electro-thermal coupling, 87, 101, 103
Error vector magnitude, 13

F

Filter, 40, 59
Frequency compensation, 35, 36, 43
Frequency tuning, 25

M. Onabajo and J. Silva-Martinez, *Analog Circuit Design for Process Variation-Resilient Systems-on-a-Chip*, DOI: 10.1007/978-1-4614-2296-9, © Springer Science+Business Media New York 2012